Chemistry

THE STUDY OF MATTER
FROM A CHRISTIAN
WORLDVIEW

Dr. Dennis Englin

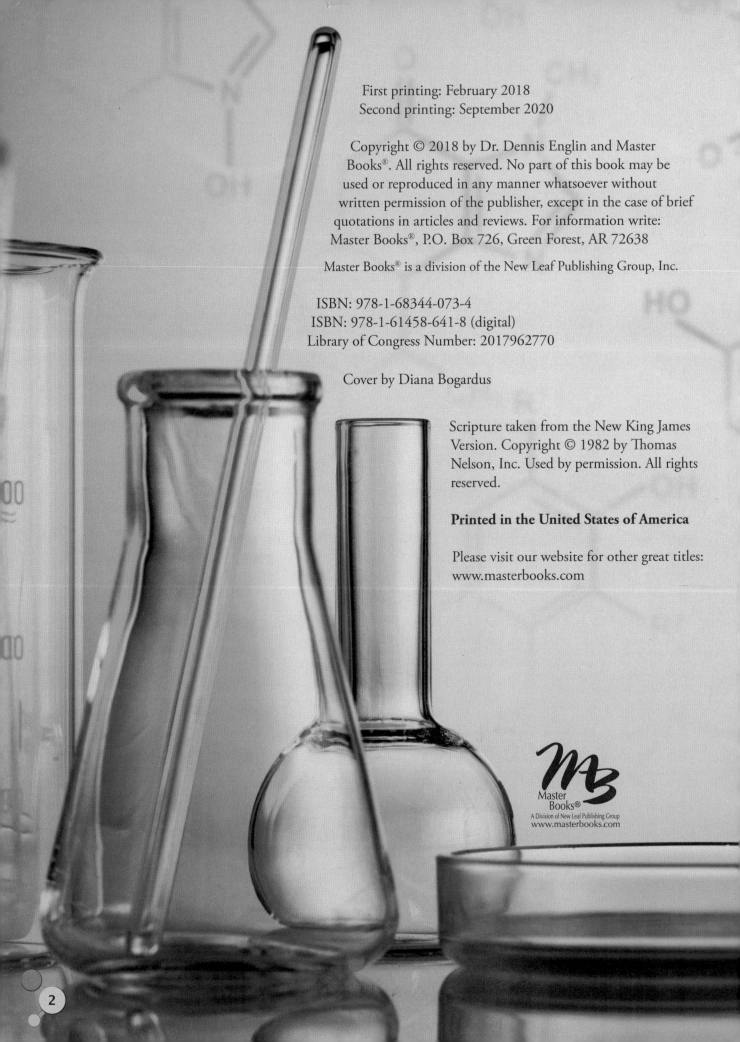

First printing: February 2018
Second printing: September 2020

Master Books® is a division of the New Leaf Publishing Group, Inc.

ISBN: 978-1-68344-073-4
ISBN: 978-1-61458-641-8 (digital)
Library of Congress Number: 2017962770

Cover by Diana Bogardus

Printed in the United States of America

Please visit our website for other great titles:
www.masterbooks.com

Master Books®
A Division of New Leaf Publishing Group
www.masterbooks.com

PRAISE FOR DR. ENGLIN'S CHEMISTRY COURSE

I was a student in Dr. Dennis Englin's high school chemistry class for homeschool students. I found the curriculum easy to understand and thoughtfully organized. The concepts I learned made my transition to university-level chemistry seamless, as every major topic had already been thoroughly covered by Dr. Englin. As a graduate from a pre-health program and now a biology professor, I would heartily recommend this curriculum to anyone looking to jump-start their undergraduate career.

> ~ Christopher Craig Chambers, M.S. computer science, California State University

Presented from a thoroughly and unashamedly biblical perspective, Dr. Englin's curriculum engages with both the fundamental basis and practical application of chemistry, all the while highlighting how it showcases the orderliness and intricacy of God's created world. The curriculum's principal content and easy-to-follow organizational structure is the outgrowth of Dr. Englin's own experience in teaching high school- and college-level science classes for nearly five decades. This book's content, paired with the accompanying laboratory studies, makes for an all-inclusive and well-rounded approach to high school chemistry, sure to benefit students and parents alike, all the while magnifying the glory of the Creator.

> ~ Lee Anderson, Jr., Christian Apologist and Writer

Dr. Englin has the amazing gift of making seemingly difficult subjects easy to understand. Through his classes, science came to life and I always looked forward to coming back the next week! I appreciated the detail, hard work, and love he poured into each student; it inspired me to want to learn.

> ~ Amy Mack Dinsmoor, former student from 2006–2010

The scientific information Dr. Englin presents in his homeschool chemistry course is rooted in a biblical worldview. It is informative and readily understandable to the high school student; it stimulates and challenges. Dr. Englin provides all of the scientific information and mathematical procedures necessary for students to complete this course. The curriculum is sufficient in scope for a year's education in chemistry, and able to pique the student's curiosity so that they desire to learn more. His laboratory studies are enjoyable and instructive, complementing the coursework presented in the lecture material. I heartily recommend this course to any family seeking science curricula for a year of chemistry.

> ~ Phillip A., former student

I am the parent of three young men who have been immensely blessed by the teaching of Dr. Dennis Englin. My sons have taken all four of Dr. Englin's high school science courses, with chemistry being their favorite. Implementing the high school chemistry curriculum, written by Dr. Englin, in my homeschool was indeed a joy. This chemistry curriculum will guide users on a captivating tour through God's chemical world. The materials are written in such a way as to make them exceptionally understandable, and user-friendly to all, even those without a strong background in the sciences. The clear introduction of the concepts, build one upon another as to incorporate all material through a spiral learning approach. The daily lessons are written directly to the student, and not only are they scientifically accurate, but are often presented with awe, insight and humor. Using this phenomenal curriculum certainly made my role as a homeschool educator easier. I highly recommend it for the Christian homeschooling family!

~ Mrs. S. Anderson

Having previously taken chemistry at another school, I remember walking into Dr. Englin's class expecting boring textbooks and confusing lectures. I was delightfully surprised. Dr. Englin ignited our curiosity and helped us understand complex topics through real-world explanations. He not only taught us chemistry, but also how to learn and reason through complex problems. His sense of wonder and excitement at God's creation was contagious and I will always have fond memories of that class.

~ Wendy Mack, former student from 2006–2010

ABOUT DR. DENNIS ENGLIN

Dr. Englin enjoys teaching in the areas of animal biology, vertebrate biology, wildlife biology, organismic biology, chemistry, and astronomy. Memberships include the Creation Research Society, Southern California Academy of Sciences, Yellowstone Association, and AuSable Institute of Environmental Studies. Dr. Englin's most recent publications include a text currently used in *Principles of Biology*. His research interests are in the area of animal field studies. He was a Professor of Biology at the Center for Professional Development at The Master's University in Santa Clarita, California, until his retirement in 2018.

B.A., Westmont College
M.S., California State University, Northridge
Ed.D., University of Southern California

TABLE OF CONTENTS

Vocabulary words
On the first page of every chapter, vocabulary words are introduced, which are bolded in that chapter's text and have brief definitions found in the glossary at the back of the book. Students are encouraged to either write these out on 3 x 5 cards or to create another useful means of reviewing these throughout their course of study. Comprehension of sometimes difficult terms and concepts is very important to completing a course in chemistry or any other complex science study.

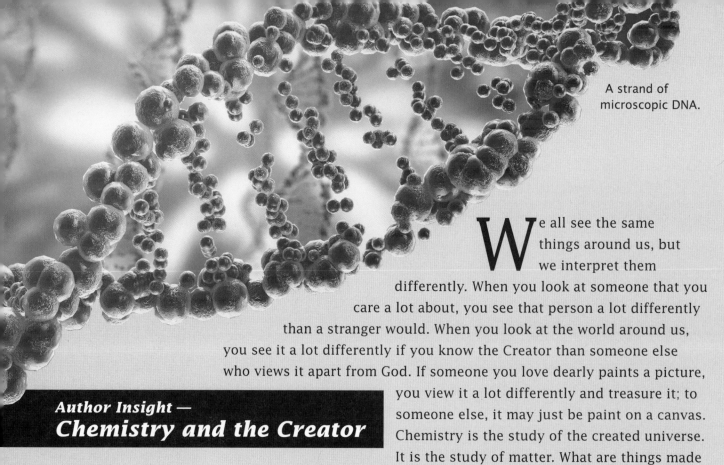
A strand of microscopic DNA.

We all see the same things around us, but we interpret them differently. When you look at someone that you care a lot about, you see that person a lot differently than a stranger would. When you look at the world around us, you see it a lot differently if you know the Creator than someone else who views it apart from God. If someone you love dearly paints a picture, you view it a lot differently and treasure it; to someone else, it may just be paint on a canvas. Chemistry is the study of the created universe. It is the study of matter. What are things made of? What holds them together? Why do some things react violently and others do nothing? To view the world around you as created by Someone that you feel is very important to you makes a difference more than if it is just particles that could not care if they bless or harm you. This is your worldview — or the lens through which you view the world.

This study is presented through the eyes of one who has walked with the Creator and Lord of the universe for many years. I pray that you will see it that way, too. Chemistry as Christ would have us see it is the goal of this study. I hope that it does not strike you as strange to think that the Holy Spirit of God can guide you through this study. He has infinite wisdom that went into the creation. We have limited understandings because we are creation. Nevertheless, He has enabled people down through the years to understand parts of His creation to show His care and love to those created in His Image. He develops our skills as we study, practice and grow. This is not a study that you can just walk through like a grassy field. Rather it has a few cliffs to climb and streams to ford. Some areas will come easier than others and some will take more time and practice. As you study each lesson, you will have to complete practice exercises and take a weekly quiz. You will conduct a laboratory procedure and write a report dealing with that week's lesson. About every 3 or 4 weeks you will stop and review and take an examination. Concepts and skills

Microchip technology can now perform chemical reactions in the search for new drug therapies.

build upon each other as you go through the study. As you go through, you will be provided with applications from Scripture as to how the material reflects the truth and mind of God in His creation. These are not just Bible verses inserted without context. They deal directly with the material studied and are a guide to keep your focus through the Mind of God.

You will see God's never changing nature and absolute control of the physical universe through the natural laws. Christ is the only explanation for the protons and electrons of atoms holding together when they should fly apart as happens in an atomic explosion. We see accountability when we violate the use of materials as I saw in high school when a teacher placed sodium metal in water and it violently exploded. We see grace in the safeguards that God gave us by revealing principles of chemistry to protect us from harm and greatly enrich our lives–like the elements making up the microchips in a computer.

As you study you are walking through the art gallery of God's wonderful creation. It is beautiful even in a fallen world. Now that is grace! The title *Chemistry Through a Christian Worldview* is to be taken seriously. A basic principle of the Christian walk is that God never leads you down useless pathways. Sometimes we take detours in our sin but He brings us back. You are in this study by divine appointment. You and I do not know how He is going to use it in your life but He certainly does.

Note: The *Master's Class Chemistry Teacher Guide* includes worksheets with weekly assignments, lab charts, and lab journaling pages, as well as answer keys for worksheets, quizzes, labs, and exams.

A violent reaction when sodium (Na) metal is added to water.

CHAPTER 1

INTRODUCTION

OBJECTIVES AND VOCABULARY

At the conclusion of this lesson the student should have an understanding (as evidenced by successfully completing the quiz at the end of this lesson) of:

1. The concepts of chemistry, scientific models, atoms, and molecules

2. The meanings of the terms mass, weight, density, chemical reactions, hypotheses, and theories

3. The differences between the sources of knowledge of chemistry and the Scriptures.

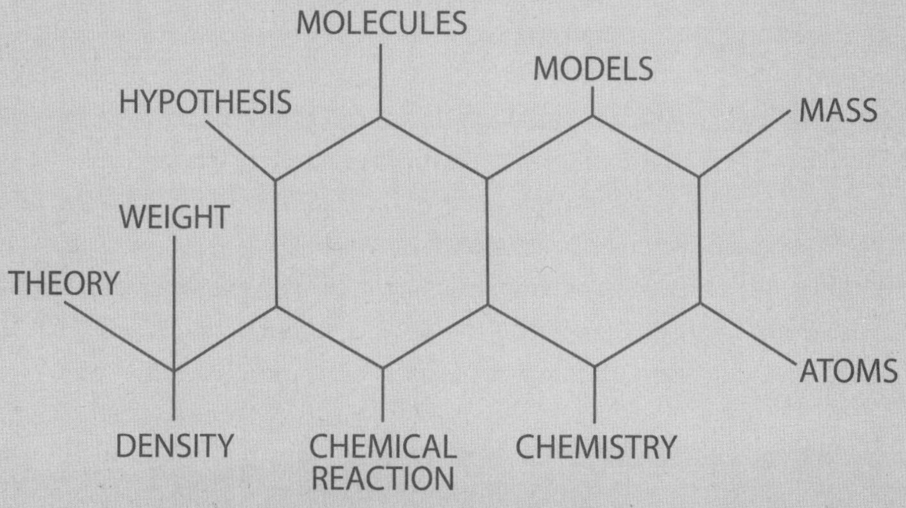

MOLECULES

MODELS

HYPOTHESIS

MASS

WEIGHT

THEORY

ATOMS

DENSITY

CHEMICAL REACTION

CHEMISTRY

Chemistry is the study of matter, which is the "stuff" that we are made of and everything around us. Matter can be broken down into very small parts that we cannot see called **atoms**. In science, we use **models** to describe what we cannot see but whose effects we can see. Table salt is composed of **molecules,** each of which are composed of one atom of sodium and one atom of chloride. We cannot see these molecules but we can taste them when we have enough of them together in one place. We describe its effects upon the taste buds of our tongues. We describe the sodium chloride molecule (NaCl) with a model of what we call a molecule. We do not know what it really looks like, but we do know its effects.

If something has a foul odor and we cannot see it, we can still smell it. I hope we cannot see it. We know it is there by its odor. The particles of the substance that goes into our noses are too small to be seen, but we still know that they are there. This is especially true when a skunk goes by. When we try to measure these particles, we have to have a lot of them. The units of measurement that we use in chemistry have this idea in mind. We cannot weigh one NaCl (sodium chloride) molecule, but we can weigh a large quantity of them.

The amount of matter in a sample is its **mass** (the mass of all of its atoms added together). The pull of gravity on the mass is its **weight**. Can the weight of an object change without changing its mass? Yes it can. If you go into orbit around the Earth in the International Space Station, your mass remains the same but your weight becomes much less. A helpful application of this concept is that if you know the mass of a sample and the mass of each of the atoms in the sample, you can find out how many atoms are in the sample. We will do this in later lessons.

Early studies of atoms demonstrated that many substances are made up of larger (but still very small) particles called molecules. An example is table sugar (sucrose) that is composed of 12 carbon atoms, 11 oxygen atoms, and 22 hydrogen atoms bonded together. When they come apart and are put back together in a different combination, a **chemical reaction** has occurred.

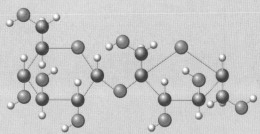

A structural formula of the chemical compound sucrose (table sugar).

We describe the size of an object by its volume, which is the amount of space it occupies.

Have you ever woken up in the middle of the night wondering why ice floats? If it did not, fish in northern lakes would die in the winter when the water froze from the bottom up. But unlike almost all other molecules, God created water molecules so that they would move farther apart from each other as water freezes into a crystal that we call ice. When the water molecules move farther apart from each other as water freezes, there are fewer water molecules in the same volume. This means that a cubic inch of ice has fewer water molecules than a cubic inch of liquid water. Therefore, a cubic inch of ice has less mass (and weight) than a cubic inch of liquid water. The mass of matter divided by its volume is its **density**. In this situation, ice has a lower density than liquid water.

Try this experiment. Take two equal volumes of water (½ cup) and dissolve about ½ of a teaspoon of table salt in one of them. Place them both in the freezer and see which one freezes first. Try to explain what happened by thinking about what happens when the salt molecules get in between the water molecules as they are trying to form ice crystals.

Your suggested explanation is called a **hypothesis**. Science is based upon making observations and proposing hypotheses. If a hypothesis is supported by many other observations and experiments, it becomes a **theory**. This is very different from a fact because a hypothesis or theory will be later replaced when someone comes up with a better hypothesis that does a better job of explaining the observations. Facts and truth do not change. A hypothesis and theory are

Facts and truth do not change. A hypothesis and theory are temporary because they are stages in our learning process.

temporary because they are stages in our learning process. This is the difference between science and Scriptures of the Christian Bible. The Bible was given to us by God sharing His wisdom with us. Science is a gift that God gave us to better understand the world that He made. Science is good because it gives us many blessings if it is not used for evil purposes.

The process of observation, hypothesis, testing the hypothesis and theory is called the scientific method. In some ways, this is what we do all the time. If you got up one morning and found a horse in your kitchen, you would start asking some questions. Why is the horse there and where did it come from and did it eat my breakfast? These are your observations. When you came up with possible explanations, they would be your hypotheses. Then your neighbor from down the street comes up and asks if you saw his horse who snuck off. Now you have information to test your hypotheses. In the scientific community, when a hypothesis is proposed it is published in a scientific journal and read by many interested in that topic. Many will test your hypothesis and if it stands up well with their further testing, it becomes recognized as a theory. Later, someone proposes a better hypothesis and everyone proclaims it to be the theory that replaces yours. Here today, gone tomorrow. Science is good but it is a growth process. Just consider how computers have changed in the last ten years. It is humans trying to understand the physical world.

In science, it is usually emphasized that the observations and experiments be reproducible. A criticism sometimes made against the idea of creation is that you cannot reproduce it and take a closer look at it. But that is also the case in forensics (the study of a crime scene) and the discussions of the evolution of life. It is, however, in science more reliable if you can reproduce the procedures and get the same results.

It is emphasized in science, as well, that your conclusions be falsifiable. For example, you cannot show a hypothesis to be true, because you cannot think of every possible hypothesis to test. Sometime later someone will come up another hypothesis that fits the data much better and replaces all earlier hypotheses. But you can show a hypothesis to be false. A number of years ago a group in Utah claimed to have developed cold fusion. That means that they claimed to be able to fuse the nuclei of two atoms of hydrogen to make a heavier atom with the release of nuclear energy that could be used to heat water into steam to drive an electric turbine. Theoretically and practically the procedure takes a lot of heat energy to slam the atoms together. That is the basis for a hydrogen bomb. When asked to duplicate their results they could not do it. It was determined that their hypothesis was false. Some state that creation cannot be included as an alternative to evolution because it is not reproducible nor falsifiable. That is true. Just like evolution, creation is neither reproducible nor falsifiable. But to reject creation on that basis is to reject God Himself, and the natural world He created. We study God through the Scriptures that He gave us to come to understand Him.

We do not use chemistry to study God but we can see evidence of God in chemistry.

We do not use chemistry to study God but we can see evidence of God in chemistry. Just the mathematical relationships in chemistry attest to the intricate design in creation. As well, the consistency of the structures of the elements and their properties testify of His superior intelligence. Oxygen behaves like oxygen no matter where it comes from. These Laws of Nature are the foundation that has to be there before we can proceed any further in our study. This is something we should not take for granted.

In Scripture, God is not testing and refining His thoughts. He had full knowledge and understanding before the beginning of creation. Our understanding of Scripture improves as we mature and grow closer to Christ but the Scripture itself never has to mature or change. Many stumble when they confuse the authority of science with that of Scripture. Some say that science is something that we can see and test and Scripture is not, so they place their faith in sight rather than God. Our study of science is good but our level of knowledge and understanding is much less than God's. As well, with science we can only study the physical reality. God was before there was a physical reality and life. We have to seek spiritual answers for the origin of physical reality. Before matter was created, there was God, so we have to go to Him and He instructs us through the Scriptures for us to understand the origin of the physical universe. God designed and created matter so when you struggle to understand chemistry, ask Him to come along side of you and be your guide and super Tutor. Wow!

It is also important to distinguish between observation and measurements and hypotheses and theories. I can measure the length of a table or the mass of a sample of

It is also important to distinguish between observation and measurements and hypotheses and theories.

NaCl. But if I talk about the structure of the NaCl molecule, I am dealing with a hypothesis about something that I cannot see. Observations and measurements are more like facts, while hypotheses and theories expound on them and provide explanations for the behavior that was observed. Hypotheses can be thought of as tentative explanations, and theories are explanations based on multiple sources with strong evidence. Interesting that some will reject the God that they cannot see but accept the description and reality of a NaCl molecule that they also cannot see.

God created everything we can see and the invisible structures within them that we cannot see. And we are still only beginning to understand how intricate, amazing, and purposeful the Creator's designs truly are!

SCIENTIFIC MODELS

REQUIRED MATERIALS

- Small box

- Random item that fits in the box

INTRODUCTION

A scientific model is a description of the behavior of something that you have no means of ever seeing with current technologies. You cannot see an atom but you can see the effects of many atoms. You cannot see a proton and you cannot see an electron. But there is something there that is identified by these names. This is a difficult concept because it is contrary to our everyday way of thinking. Can you imagine getting into and riding in an invisible car? Kind of silly, isn't it? But that is what we do with many things in science. We have a model that is a description of something that would behave just like something that we cannot see. The model of an atom is not an actual description of what an atom looks like. It cannot be because we do not know what an atom looks like. The description is the description of something that would behave just like an atom. From this model, we can predict what an atom would do under other circumstances. Our goal is not to describe what an atom actually looks like but rather what it will do.

TEACHER NOTE

This lab requires that you prepare a small, sealed box with an object, unknown to the student, inside for them to analyze.

PURPOSE

This lab exercise is designed to demonstrate how scientific models are designed and used to understand things that cannot be directly observed such as atoms and molecules.

PROCEDURE

This lab is an exercise in constructing a scientific model. Perhaps this will give you a better idea of what a model is and its limitations. You have a sealed box. It has an object in it. You can do almost anything to your box except alter, destroy, or open it. You are not at any time to state or guess what you think is in the box. You will not be shown what is in the box. That is the way it is with atoms and their parts. You are to describe as many properties of the object as you can — but never to identify it! For example, tilt the box and determine if the object slides or rolls in the box. How fast does it roll or slide? What if you tilt it the other way? Does it respond differently? As you hold the box, does the object feel heavy?

Remember that your description cannot have anything to do with what you might think is in the box. Describe at least 6 procedures you perform with the box, your observations, and conclusions. Always use complete sentences. You are not just writing this report for yourself. One of the purposes of the laboratory reports is to improve your writing skills. Part of the grade on the report is how well you follow instructions. At the end of the report, summarize the properties that you can identify for the object in the box. Your report will also be graded on how neat and well organized it is. It can be hand written, but it must be clear.

REPORT

Scientific models

Give a unique name to the object in the box even though you do not know what it is.

For each procedure: describe the procedure and state your observations and conclusions.

Summarize the properties of the object in the box. Remember — do not try to identify what is in the box.

How do you think that this is similar to the way atoms and molecules are studied?

CHAPTER 2

METRIC MEASUREMENTS IN CHEMISTRY

OBJECTIVES AND VOCABULARY

At the conclusion of this lesson the student should have an understanding (as evidenced by successfully completing the quiz at the end of this lesson) of:

1. Common units used in the metric system

2. Units of density and how density is used to explain flotation patterns

3. Application of the prefixes milli-, centi-, deci- and kilo- in the metric system.

DENSITY

GRAM

METRIC SYSTEM

MICRO

KILO

LITER

CENTI

METER

DECI

MILLI

Measurements are very important in all of the sciences and everyday life. If you take a medication, it has to be very carefully measured out. That has not always been the case. Can you imagine some of the results of that? The use of measurements probably go back to Adam. They definitely had to be used in building the Tower of Babel. Without accurate means of measuring and planning, the tower never would have stood.

The concept of accurate measurements came from the mind of God.

"A just balance and scales are the Lord's; all the weights in the bag are His work." (Proverbs 16:11)

Throughout the universe, numbers are important, such as the gravitational forces holding galaxies together and the wavelengths of light emitted from the distant edges of the universe. God's wisdom designed the forces and elements at the beginning of creation.

"I, wisdom, dwell with prudence, and I find knowledge and discretion. The Lord possessed me at the beginning of His work, the first of His acts of old. Ages ago I was set up, at the first, before the beginning of the earth... when He made firm the skies above, when He established the fountains of the deep, then I was beside Him, like a master workman, and I was daily His delight, rejoicing before Him always, rejoicing in His inhabited world and delighting in the children of men." (Proverbs 8: 12, 22, 23, 28, 30, 31)

The universe functions like well-designed clock work. All of the forces in the universe are balanced mathematically to keep every galaxy, star, planet, moon, and molecule in perfect positions in relation to each other to give us a beautiful place to live. These forces are seen as well in the diverse elements making up the universe, our world and bodies. God chose to share this wisdom with us who are created in His image. He enables us to see the wisdom of His handiwork when our eyes are opened by our redemption from the blinding effects of sin through faith in Jesus Christ.

In the 1600s most of the early scientists that laid the foundations of modern science were born-again Christians. When you can stand back and see the patterns of the nature of the elements, you cannot help worshipping the Creator who spoke it all into being showing us the wisdom of His mind.

In 1791, Antoine Lavoisier (called the Father of Modern Chemistry) helped develop the metric system. He also developed an extensive list of elements and made major contributions to the naming of the elements and compounds. Prior to this different countries had different measuring systems. For example, different countries had different lengths for measuring a foot. This was very confusing. Unfortunately, Lavoisier was a member of the French aristocracy and involved in designing the system of taxation right before the French Revolution. He was one of the first ones to go to the guillotine.

The metric system is called the Systeme International d' Unites (SI system) or International System of Units.

The **metric system** is used in scientific measurements because everything comes in units of tens. In the English system there are different numbers of units used such as 12 inches in a foot and 3 feet in a yard.

Lavoisier by Jacques-Léonard Maillet, ca. 1853.

The standard unit for length in the metric system is the **meter** (about 39 inches), the standard unit of volume is the **liter** (about 1.06 quarts), and the standard unit of mass is the **gram**.

The size of a unit of measurement is known by its prefix. **Deci-** means 1/10 (one tenth). **Centi-** means 1/100 (one hundredth). **Milli-** means 1/1,000 (one thousandth). **Micro-** means 1/1,000,000 (one millionth) and **Kilo-** means 1,000 (one thousand).

Prefix	Numerical Meaning
micro (μ)	0.000001
milli (m)	0.001
centi (c)	0.01
deci (d)	0.1
	1
deca (D)	10
hecto (H)	100
kilo (k)	1,000

A decimeter (dm) is .1 m (m stands for meter and dm stands for decimeter). A centimeter (cm) is .01 m and a millimeter (mm) is .001 m. A kilometer (Km) is 1,000 m.

A deciliter (dl) is .1 L (L and l stand for liter). A centiliter (cl) is .01 l and a milliliter (ml) is .001 l. Kiloliters are not usually used in chemistry. A thousand liters is a lot.

A decigram (dg) is .1 g (g stands for gram). A centigram (cg) is .01 g. A milligram (mg) is .001g and a kilogram (Kg) is 1,000 g.

meter (m)
length

liter (l)
volume

gram (g)
mass

In the metric system, the standard unit of measurement for weight (force of gravity) is the Newton — named after Isaac Newton who expressed the Laws of Motion. In chemistry, when you weigh something on a balance the units used are usually milligrams. But milligrams are not weight. The balance is actually measuring the weight but the scale of the balance is adjusted to give mass (milligrams). Milligrams are used in chemistry because of the small quantities of materials used.

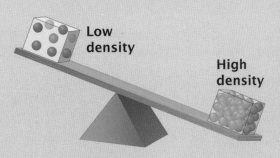

Low density

High density

Comparing the mass of equal volumes demonstrates the density of the two objects.

The **density** of an object is measured as mass divided by volume, which is g/ml (the number of grams making up each milliliter). In chemistry the units of grams and milliliters are more often used because small quantities are used. A ml of iron is much heavier than a ml of Jell-O® because the density of iron is much greater than that of Jell-O®. Ice floats in liquid water because the density of ice is less than that of liquid water. This is also why a several-hundred-ton aircraft carrier floats in water. Most of the volume of the aircraft carrier is air, which is much lighter than water. The total weight of the aircraft carrier is less than the same volume of water so the aircraft carrier floats. When the king of Sweden had a war ship (*Vasa*) built, he insisted that they put more bronze cannons on the ship. When they launched the ship for the first time, it sank just outside of the harbor in 1628. It was later salvaged in 1961 and placed in the Vasa Museum in Stockholm, Sweden, in 1988.

REACTIONS IN ACTION

Do the following exercise to test your understanding of density. Liquid A has a density of 1.05 g/ml; liquid B has a density of 1.10 g/ml; liquid C has a density of 0.97 g/ml; and liquid D has a density of 1.00 g/ml. When all 4 liquids are poured together into a tall glass, they do not mix; instead they settle out forming 4 layers.

Liquid A	1.05 g/ml
Liquid B	1.10 g/ml
Liquid C	0.97 g/ml
Liquid D	1.00 g/ml

Which liquid will be on the bottom? Which one will be second, floating on the bottom layer? Which liquid will be third up from the bottom and which one will be on top?

Form a hypothesis (a principle or explanation that you make from your observations) explaining how you came up with your answer. Afterwards, and not before, compare your answer to the one below.

Answer: the bottom layer is B because it has the greatest density (1.10 g/ml). On top of B is liquid A (1.05 g/ml). On top of A is liquid D (1.00 g/ml) and liquid C is the top layer with the lowest density (0.97 g/ml). How did you do?

LABORATORY 2

THE METRIC SYSTEM

REQUIRED MATERIALS

- Square or rectangular object

- Ruler with measurements in inches

- Graduated cylinder (10 ml)

- Weighing boat

- Scale

INTRODUCTION

The metric system was developed on the basis of the number 10. The following prefixes are commonly used:

- *milli* meaning 1 thousandth

- *centi* meaning 1 hundredth

- *deci* meaning 1 tenth

- *kilo* meaning 1 thousand

The meter is the metric system's unit of length. It is equivalent to about 39 inches in the English system. By the list above: a decimeter is a _tenth_ of a meter, a centimeter is a _hundredth_ of a meter, and a millimeter is a _thousandth_ of a meter. A kilometer is _thousand_ meters.

See the chart below for the units for length, volume, mass, and force in both the metric and English systems.

Quantity	Metric Unit	English Unit
length	meter	foot (12 inches)
volume	liter	gallon
mass	gram	slug (32.174 lb)
force (weight)	newton	pound

PURPOSE

This exercise is designed to familiarize you with the use of metric units. The sciences exclusively use metric standards of measurement.

PROCEDURE

You will find the following English–metric conversions helpful for this exercise.

English	Metric
1 inch	= 2.54 cm (centimeters)
1 gallon	= 3.8 liters
1 mile	= 1.61 km (kilometers)
1 pint	= 0.473 liter

1. Solve the following:

   A. If a gas station charges $1.25 for a liter of gasoline, how much is it for a gallon of gasoline? $4.75 gallon

   B. If you travel 10 miles, how many kilometers have you traveled? 16.1 kilo

   C. If you have 3 pints of fruit juice, how many liters do you have? 1.419 leters

2. Find a square or rectangular object.

   A. Measure its length, width, and height in inches.

   B. What is its volume in cubic inches? 7.3 in³ (length x height x depth)

   C. Convert each of the measurements into cm.
      7.3 in³ x 2.54 x 3 = 1,168 cm³
   D. What is the volume of the object in cubic centimeters (cc)?

   E. A cubic centimeter is exactly the same as a milliliter (ml). What is the object's volume in milliliters and liters?

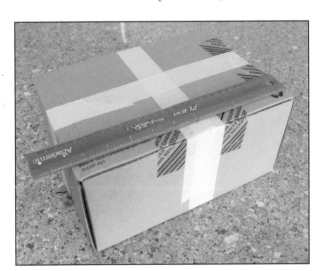

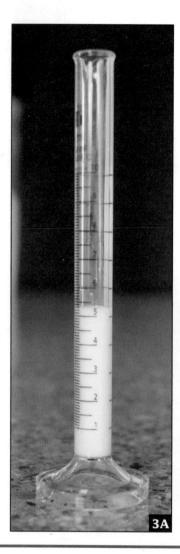

3. Mass / Density Measurements

A. Measure out 5 ml of a liquid other than water with a 10 ml graduated cylinder. Line up the middle, not the edge, of the surface of the liquid (the meniscus) with the markings on the graduated cylinder.

B. Place a weighing boat on a scale and find its mass in grams.

C. Add the 5 ml of liquid to the weighing boat and find the mass of the 5 ml of liquid including the mass of the weighing boat.

D. Calculate the mass of the liquid by subtracting the mass of the weighing boat from the combined mass.

E. Divide the mass of the liquid in g by the volume in ml. This gives you the density of the liquid in units of g/ml.

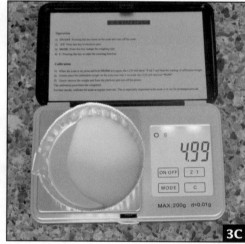

LAB INSIGHTS

In chemistry, volumes of liquids are typically measured with graduated cylinders, burettes, pipets, and volumetric flasks.

The graduated cylinders measure volumes as small as ml (milliliter); pipets can measure volumes as small as 0.1 ml; and a volumetric flask can only measure a set volume — such as 100 ml.

Left to right: graduated cylinders, pipet, and volumetric flask, and burette (not to scale).

SIGNIFICANT FIGURES

The level of measurement made with a tool in chemistry determines the level of accuracy (how close you are to the true value) of any calculations using the measurement. For example, if three people each measure the volume of the same liquid and get 50. ml, 49. ml, and 50. ml, you cannot express the average as 49.7 ml because the graduated cylinders cannot measure a tenth of a ml. If your measurements were made with pipets instead (which can measure a tenth of a ml), the answer 49.7 ml would be fine because the measurements in a pipet would be expressed as 50.0, 49.0, and 50.0. In the answer 49.7 ml, when based off the 50., 49., and 50. numbers, the numbers 49. would be called the significant figures.

If you used a scale that measures down to a tenth of a gram and you got the measurements of 30.2 g, 30.0 g, 31.0 g, and 30.9 g for the same sample, the average would be 30.525 g., but the scale can only measure down to the level of a tenth of a gram, so the answer needs to be rounded off to 30.5. The trailing 25 is uncertain and the significant figures are 30.5.

Suppose you measured the following volumes of HCl (hydrochloric acid) that reacted with another substance: 20.5 ml, 25.4 ml, 22.7 ml, and 24.2 ml. The average of these results is 23.2 ml. This number is fine because the last number is a tenth of a ml and that is the limit of the measurement of the burette.

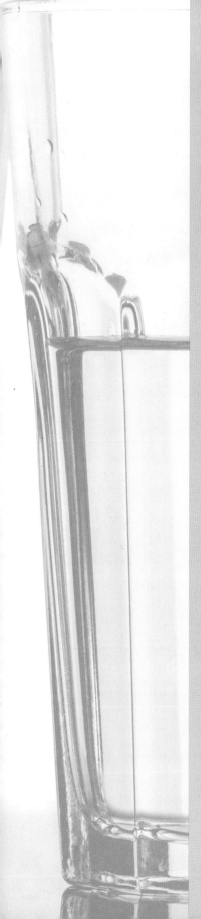

CHAPTER 3

CHEMICAL SOLUTIONS – PERCENT CONCENTRATIONS

OBJECTIVES AND VOCABULARY

At the conclusion of this lesson the student should have an understanding (as evidenced by successfully completing the quiz at the end of this lesson) of:

1. The nature of a solution as being a solute dissolved into a solvent

2. How to determine the amount of solute and solvent to use to prepare a solution of known percent concentration.

CONCENTRATION

SOLVENT

SOLUBILITY

SOLUTION

PERCENT
CONCENTRATION

SOLUTE

A solution is a mixture. We normally think of a solution as something dissolved in water. But it could be something dissolved in oil as well.

Imagine that you have 4 glasses that appear to each contain water. The first has just water; the second has water and salt; the third has water and sugar; and the fourth has water and quinine (very bitter). They all look alike. Does that mean that they really are alike? Appearance may not always be the best judge. Suppose you taste each (which you never do in a chemistry lab). Now are they all alike? Obviously not. What if you have 2 clear glasses of water and one has a little salt and the other has a lot of salt? Could you tell the difference? It would be quite obvious that what a liquid contains and how much it contains are both very important. Along the same line of thought, have you ever taken a large drink of pickle juice?

An easy way to describe a solution is that a **solute** is dissolved into a **solvent** to make a **solution**. Most substances studied in chemistry are dissolved in water. This makes water the solvent. When you prepare a salt water solution, salt is the solute, water is the solvent, and salt water is the solution. Some important solutions include sea water, fresh water, rain water, and the body fluids that bathe your cells.

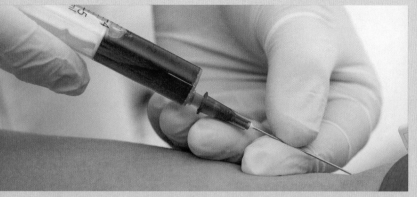

How well something dissolves is called its **solubility.** The amount of solute divided by the amount of solution is the **concentration**. When you go to a doctor's office they often take a blood sample. The concentrations of several solutes in your blood are used to determine the health of your body and cells. This is a lot better than going after you with a scalpel.

A common method of indicating concentrations is the **percent concentration.** This is found by dividing the mass of the solute by the total mass of the solution and multiplying it by 100 to get a percent.

REACTIONS IN ACTION

If you dissolve 10 grams of NaCl (sodium chloride, table salt) in water to make 100 grams of solution, you get a 10 percent NaCl solution.

10 divided by 100 is 0.1.
0.1 multiplied by 100 is 10 percent.

If you add 10 grams of NaCl to 90 grams of water, you will have a total of 100 grams of solution.

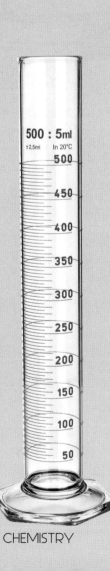

It is easier to measure out NaCl (a solid) on a balance than water, which is a liquid. One gram of water at room temperature is very close to the mass of one milliliter of water. So, instead of measuring out 90 grams of water on a balance, it is easier to measure out 90 ml of water in a graduated cylinder.

To prepare the solution, measure out the 90 ml of water and add the 10 grams of NaCl to it. Even though you only get a volume of 90 ml of solution, it is 90 ml of a 10 percent NaCl solution.

Study this example.

How would you prepare 90 ml of a 10 percent sucrose (table sugar) solution? Add 10 grams of sucrose to 90 ml of water. This gives 100 grams of solution and 10 out of the 100 (10 percent) is sucrose.

Add 10 grams of sucrose to 90 ml of water. 10 grams of sucrose and 90 grams (ml) of water give 100 grams of solution.

$$\frac{10}{100} \times 100\% = 0.1 \times 100\% = 10\%$$

How would you prepare a 20 percent solution of NaCl?

Add 20 grams of NaCl to 80 ml (grams) of water to make 100 grams of 20 percent NaCl solution. 20 grams of NaCl and 80 grams of water equal 100 grams of solution.

$$\frac{20}{100} \times 100\% = 0.2 \times 100\% = 20\%$$

If you need more than 80 ml of 20 percent NaCl solution, double the amounts. Add 40 grams of NaCl to 160 ml (grams) of water to make 200 grams of NaCl solution.

$$\frac{40}{200} \times 100\% = 0.2 \times 100\% = 20\%$$

NON-SOLUBLE LIQUID

LAB INSIGHTS

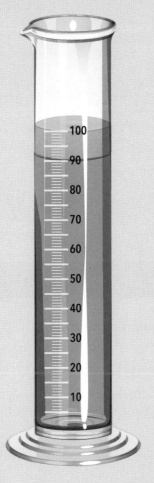

The oil represents 10 percent of the total liquid in the graduated cylinder.

LABORATORY 3

PREPARING PERCENT CONCENTRATION SOLUTIONS

REQUIRED MATERIALS

- Graduated cylinder (10 ml)

- Stirring rod

- Weighing boats / weigh paper

- Scale

- Table salt

- Sugar

- Beaker (100 ml)

- Beaker (250 ml)

- Laboratory scoop

INTRODUCTION

Percent concentrations are not difficult to understand but can be challenging to prepare. Percent concentrations are used more often in medical or physiological applications rather than in a chemistry lab. If you do not pursue chemistry further in college, you will use percent concentrations and probably not use molar concentrations.

PURPOSE

The purpose of this lab exercise is to gain experience with percent concentrations and to get comfortable with their preparation.

PROCEDURE

1. Water & Salt (Part 1)

   A. Measure out 9 ml of water with a graduated cylinder.

      (1 ml of water has a mass of 1 gram, so 9 ml of water has a mass of 9 grams.)

   B. Place the weighing boat onto the scale. (Important: See "Weighing" in Appendix 1 on page 272)

   C. Measure out 1g of NaCl (see "Weighing" in Appendix 1)

   D. Pour the water into a 100 ml beaker and dissolve the 1 g of NaCl into it. Use a stirring rod to help dissolve the NaCl.

   E. What is the percent concentration of the solution? Remember that the percent concentration is the mass of solute (NaCl) divided by the mass of the solution (NaCl and H2O) x 100 percent.

   F. How would you prepare twice the volume of the same percent concentration solution of NaCl and water?

2. Water & Salt (Part 2)

    A. Measure out 16 ml of water and pour it into a 100 ml beaker.

    B. Measure out 4 g of NaCl and dissolve it into the 16 ml of water.

    C. What is the percent concentration of NaCl in your solution

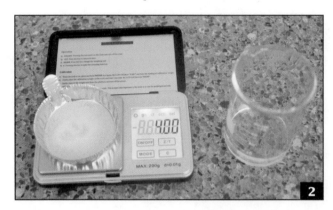

3. How would you prepare 100 ml of a 5 percent sucrose (table sugar) solution? Prepare this solution.

4. Determine the solubility of NaCl in water.
   Solubility is defined as the maximum amount of a solute that can dissolve into a given volume of solvent.

    A. Measure out 100 ml of water using a graduated cylinder and pour it into a 250 ml beaker.

    B. Fill a weigh boat approximately half full with NaCl and record the mass.

    C. Gradually add the NaCl to the 100 ml of water while stirring until no more will dissolve.

       You may need to refill the weigh boat with additional NaCl to be added to the water. If you do, keep track of the mass of the additional NaCl used.

    D. Determine the maximum number of grams of NaCl that will dissolve in 100 ml of water.

Calculate the percent concentration of NaCl as

$$\frac{\text{the number of grams of NaCl}}{\text{(grams of NaCl + water)}} \times 100\%$$

For example, if you dissolve 23 g of NaCl into 100 ml of water and no more dissolves, the **solubility** is

$$\frac{23 \text{ g NaCl}}{100 \text{ ml } H_2O}$$

The **percent concentration** is

$$\frac{23 \text{ g NaCl}}{(23 \text{ g NaCl} + 100 \text{ g } H_2O)} \times 100\% = \left(\frac{23}{123}\right) \times 100\% = 19\%$$

The lab report should include a description of all your procedures and the answers to your calculations. Be sure to always show your work. This way it is evident that you have a grasp of the concepts even if you made an arithmetic error somewhere.

NON-SOLUBLE VS SOLUBLE

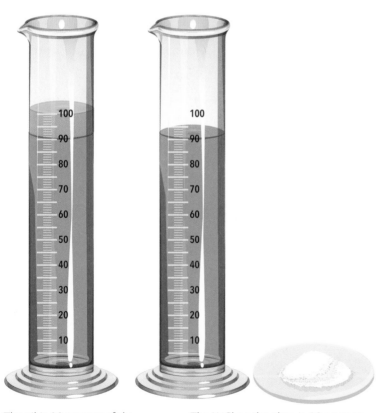

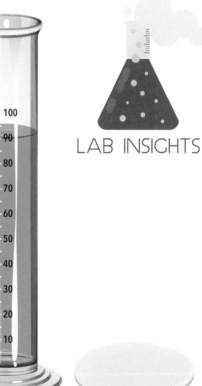

LAB INSIGHTS

The oil is 10 percent of the total liquid in the graduated cylinder, and the oil is not soluble in the water.

The NaCl on the plate is 10 percent of the total solution.

When the NaCl is poured into the water, it dissolves because it is soluble in water.

CHAPTER 4

CHEMICAL SOLUTIONS – MOLARITY

OBJECTIVES AND VOCABULARY

At the conclusion of this lesson the student should have an understanding (as evidenced by successfully completing the quiz at the end of this lesson) of:

1. The meanings of the terms atomic mass, molecular mass, Avogadro's number, mole, and molarity

2. How to convert grams of solute to moles of solute

3. How to prepare a liter of solution of known molarity

4. How to prepare different volumes of solutions of known molarity.

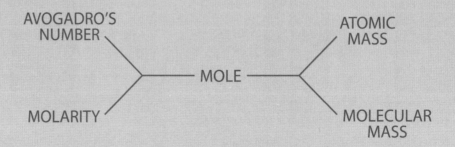

Another very common method of indicating the solution concentration in a chemistry lab is the **molarity**. When using molarity, you are indicating how many molecules of solute there are in 1 liter of solution. Percent concentration does not give the number of molecules of solute in your solution. Of course, as you realize, you cannot count the number of sugar molecules. You can, however, count out the number of sugar molecules that you place into a solution if you do a lot of them at one time.

A rather clever way of counting molecules involves the use of what is called **Avogadro's number**. This is like a dozen but a much larger number. It is okay to count out a dozen bananas or doughnuts–but not a dozen molecules because they are simply too small. This is where Avogadro's number comes in handy. Just as a dozen is 12, Avogadro's number is 6.022×10^{23}. This is 6,022 with 20 more zeros after it. The number is so large that this many kernels of unpopped popcorn would cover the United States to a depth of 9 miles.

$$602,200,000,000,000,000,000,000$$

Avogadro said that molecules are made up of atoms and that gases (such as air) with the same volume and temperature have the same number of gas molecules. Several others later used Avogadro's findings (after his death) to come up with Avogadro's number, so it was named after him. Instead of using the word dozen to refer to a number that large, they chose the word **mole** (not the furry creature that burrows under the ground). A number this large makes good sense because atoms and molecules are so small.

Later, the concepts of atomic mass and molecular mass were developed. They realized that if you had 1 gram of hydrogen atoms, you would have 1 mole of hydrogen atoms or 6.022×10^{23} hydrogen atoms. Some call this the atomic weight of hydrogen (grams are actually units of mass). It is usually called the **atomic mass**. It was also found that 1 mole of oxygen atoms has a mass of 16 grams. So if you measured out 16 grams of oxygen atoms, you would have 6.022×10^{23} oxygen atoms. This is what is meant by saying that you cannot measure out an oxygen atom but you can measure out a mole of oxygen atoms.

By knowing that 1 mole of H_2O (water) molecules is made up of 2 moles of H (hydrogen) and 1 mole of O (oxygen), you can find the **molecular mass** of water.

2 moles of H	are	2 moles x 1 gram/mole	=	2 grams
1 mole of O	is	1 mole x 16 grams/mole	=	16 grams
1 mole H_2O			=	18 grams

Na (sodium) has an atomic mass of 23.0 grams/mole and Cl (chlorine) has an atomic mass of 35.5 grams/mole. NaCl has a combined molecular mass of 58.5 grams/mole (23.0 + 35.5). There is 1 atom of Na and 1 atom of Cl in NaCl so the molecular mass is just the sum of their atomic masses.

Na	= 23.0 grams/mole
Cl	= 35.5 grams/mole
NaCl	= 58.5 grams/mole

Usually in chemistry, solutes are measured in moles to make solutions. This is because it enables us to predict how much of a compound will react with another as you will see in the following lessons.

The term molarity refers to how many moles of solute are dissolved in a liter of solvent (usually water).

This means that if you want to make a 1M (1 molar) solution of NaCl, you would add 58.5 grams of NaCl to a liter of water. This is written as 1M or 1Mol/L.

$$1 \text{ M NaCl} = \frac{1 \text{ mole NaCl}}{1 \text{ liter } H_2O} = \frac{1 \text{ mole NaCl} \times \frac{58.5 \text{ grams NaCl}}{1 \text{ mole NaCl}}}{1 \text{ liter } H_2O} =$$

$$\frac{1 \text{ mole NaCl} \times \frac{58.5 \text{ grams NaCl}}{1 \text{ mole NaCl}}}{1 \text{ liter } H_2O} = \frac{58.5 \text{ grams NaCl}}{1 \text{ liter } H_2O}$$

If you want to make half as much (1/2 liter) of solution, use half as many grams of NaCl (58.5 grams/2 = 29.25 grams) and half a liter of water. It still has a molarity of 1M even though you did not make a full liter.

$$\frac{58.5 \text{ grams NaCl}}{1 \text{ liter } H_2O} \times \frac{\frac{1}{2}}{\frac{1}{2}} = \frac{29.25 \text{ grams NaCl}}{0.5 \text{ liter } H_2O}$$

REACTIONS IN ACTION

How many moles of NaCl are there in 1 liter of a 3M NaCl solution? Each liter has 3 moles of NaCl. If you wanted to make a liter of 3M NaCl, you would add 3 moles (3 x 58.5 grams(g) = 175.5 grams(g) of NaCl) to 1 liter of water.

$$3 \text{ M NaCl} = \frac{3 \text{ mole NaCl}}{1 \text{ liter } H_2O} = \frac{3 \text{ mole NaCl} \times \frac{58.5 \text{ g NaCl}}{1 \text{ mole NaCl}}}{1 \text{ liter } H_2O} =$$

$$\frac{3 \text{ mole NaCl} \times \frac{58.5 \text{ g NaCl}}{1 \text{ mole NaCl}}}{1 \text{ liter } H_2O} = \frac{3 \times 58.5 \text{ g NaCl}}{1 \text{ liter } H_2O} = \frac{175.5 \text{ g NaCl}}{1 \text{ liter } H_2O}$$

How many moles of NaCl are there in 500 ml of 3M NaCl? 1 liter of 3M NaCl contains 3 moles of NaCl, so half as much (500 ml) would have 3/2 or 1.5 moles of NaCl.

$$3 \text{ M NaCl} = \frac{3 \text{ mole NaCl}}{1 \text{ liter } H_2O} \times \frac{1}{2} \text{ liter } H_2O =$$

$$\frac{3 \text{ mole NaCl}}{1 \text{ liter } H_2O} \times \frac{1}{2} \text{ liter } H_2O = \frac{3 \text{ mole NaCl}}{1} \times \frac{1}{2} = \frac{3}{2} \text{ mole NaCl} = 1.5 \text{ mole NaCl}$$

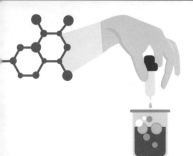

LABORATORY 4

MEASURING MOLES

REQUIRED MATERIALS

- Weigh boats

- Scale

- Laboratory scoop

- Table salt

- Sugar

- Baking soda

INTRODUCTION

The mole in chemistry is like a dozen in everyday life. The mole is used because atoms and molecules are so small. Instead of 12 objects, a mole is 6.02×10^{23} objects. That is 602 with 21 zeros after it. One big number.

$$602,000,000,000,000,000,000,000$$

PURPOSE

The purpose of this exercise is for the student to gain experience in working with the concept of the mole.

PROCEDURE

Go outside and dig up a furry creature that has been digging up your front yard and put him on the scale to find his grams. Oops! Wrong mole!

1. Moles of Table Salt

   A. Measure out 10 grams of NaCl.

   B. How many moles of NaCl do you have?

   > If you had 15 grams of NaCl, this is how you would convert it into moles. The atomic mass of Na is 23.0 g/mole and the atomic mass of Cl is 35.5 g/mole. The combined molecular mass of NaCl is (23.0 + 35.3) = 58.5 g/mole.
   >
   > $$\begin{array}{ll} \text{Na} & = 23.0 \text{ grams/mole} \\ \underline{\text{Cl}} & \underline{= 35.5 \text{ grams/mole}} \\ \text{NaCl} & = 58.5 \text{ grams/mole} \end{array}$$
   >
   > $$\frac{15 \text{ grams}}{58.5 \ \frac{\text{gram}}{\text{mole}}} = 0.26 \text{ mole}$$

2. Moles of Sucrose

   A. Measure out 10 grams of sucrose (table sugar) like you did the NaCl in procedure 1.

   B. How many moles of sucrose ($C_{12}H_{22}O_{11}$) do you have in 10 grams? Solve it like you did for the NaCl but use the molecular mass of sucrose.

   C. Look at the 10 grams of NaCl and the 10 grams of sucrose.

   Are they the same number of moles? If not, which has more moles? How is it possible for 10 grams of the NaCl and 10 grams of the sucrose to be different numbers of moles?

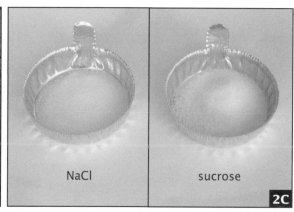

NaCl   sucrose

D. If you weighed out 100 grams of grapes and 100 grams of pineapples, would you have the same numbers of grapes and pineapples? How is this like comparing the numbers of moles of NaCl and sucrose?

3. Moles of Baking Soda

   A. Weigh out 10 grams of sodium bicarbonate ($NaHCO_3$, baking soda) like you did the NaCl and sucrose in procedures 1 and 2.

   B. How many moles of $NaHCO_3$ do you have? Do you have more or fewer moles than the moles of NaCl and sucrose? Which is the largest molecule NaCl, $C_{22}H_{22}O_{11}$, or $NaHCO_3$? Which is the smallest?

3A

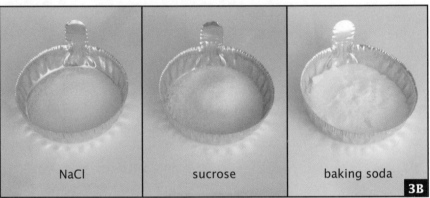

| NaCl | sucrose | baking soda |

3B

In 1808, Joseph Louis Gay-Lussac observed that at constant temperature and pressure, the volumes of gases that reacted with each other were in ratios of small whole numbers. He found that 2 equal volumes of hydrogen react with 1 equal volume of oxygen to form 2 equal volumes of water vapor. In 1811, Amedeo Avogadro interpreted Gay-Lussac's results to be that equal volumes of gas at the same temperature and pressure contained equal numbers of molecules.

2 volumes of hydrogen   +   1 volume of oxygen   →   2 volumes of water
(H_2)   (O_2)   (H_2O)

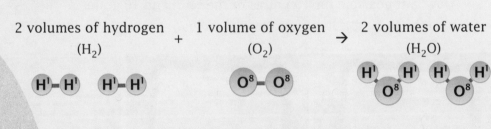

From these relationships, Avogadro determined the correct formula of water to be H_2O and not HO.

Joseph Louis Gay-Lussac

DIMENSION ANALYSIS

On page 39 you have

$$\frac{3 \text{ mole NaCl}}{1 \text{ liter } H_2O} \times \frac{1}{2} \text{ liter } H_2O = \frac{3 \text{ mole NaCl}}{1} \times \frac{1}{2} = \frac{3}{2} \text{ mole NaCl} = 1.5 \text{ mole NaCl}$$

notice the liter H_2O is crossed out. What happened there? Consider

$$\frac{1}{2} \times 2 = 1$$

Here you can see 2 divided by 2 is 1. That is the same as crossing out the 2's because one of them is in the denominator and the other is in the numerator.

This also can be done with units. If you have 30 g of NaCl, do you multiply it by 58.5 g/mole or divide it by 58.5 g/mole to get the number of moles? This is from the problem on page 39.

Dimension analysis is when you look at the units without the numbers to see how to solve the problem.

If you multiply grams by grams/mole (g x g/mole), you get g^2/mole, which is not what you want.

$$g \times \frac{g}{\text{mole}} = \frac{g \times g}{\text{mole}} = \frac{g^2}{\text{mole}}$$

However, if instead you divide grams by grams/mole you get g/g which cancels each other, giving 1 and 1/(1/mole), which is mole.

$$\frac{g}{\frac{g}{\text{mole}}} = g \times \frac{\text{mole}}{g} = \text{mole}$$

So, you solve the problem by dividing grams by grams/mole to get moles.

Now plug in the numbers and get ...

$$\frac{30 \text{ g}}{\frac{58.5 \text{ g}}{\text{mole}}} = 30 \text{ g} \times \frac{\text{mole}}{58.5 \text{ g}} = 0.5 \text{ mole (rounded off)}$$

Remember — take the numbers out and play with the units until you get the right combinations and then plug in the numbers and crank out the right answer.

CHAPTER 5

MOLECULAR MASS AND ATOMIC THEORY

OBJECTIVES AND VOCABULARY

At the conclusion of this lesson the student should have an understanding of the following as evidenced by successfully completing the chapter quiz:

1. The properties of protons, electrons, and neutrons

2. The interactions between negative and positive charges

3. How to determine the atomic mass of an isotope, knowing the number of protons and neutrons

4. How to describe the isotopes of hydrogen

5. How to determine the molecular masses of molecules given their chemical formulae.

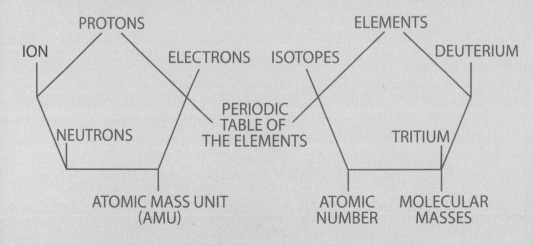

Over two hundred years ago, the English school teacher, John Dalton, after studying the writings of many other scientists, presented favorable arguments for what became known as the Atomic Theory. He concluded that an atom was the smallest particle of matter. In the latter part of the 1800s, **electrons** and **protons** were discovered to be smaller parts of atoms. It was not until 1932 that James Chadwick discovered **neutrons** as additional parts of atoms. Since then over a hundred smaller particles were found that make up protons and neutrons.

Even though the physical universe is not infinite, we see so much detail on the smaller atomic scale and the larger scale of the universe that we will never be able to see or describe it all. What is conceivable to God seems infinite to us. His understanding is far greater than ours. This is why we must always remember that His wisdom that He has shared with us in the Scriptures is greater than we will discover through our studies of physical reality.

Bronze statue of the chemist John Dalton by William Theed that stands on Chester Street in Manchester.

The British scientist H.G. Moseley discovered that atoms of the same **element** always have the same number of protons. For example, every hydrogen atom that he studied had only one proton. Every helium atom was found to have two protons and every carbon atom had six protons. Moseley called the number of protons the **atomic number**. This demonstrated a very well-organized, consistent universe as created by God's deliberate plan. Numbers are part of everything created.

Mathematics was created as part of the wisdom that God brought forth before creation as described in the Book of Proverbs.

"The Lord possessed me (wisdom) at the beginning of His work, the first of His acts of old. ... When He established the heavens, I was there; when He drew a circle on the face of the deep, when He made firm the skies above, when He established the fountains of the deep, when He assigned to the sea its limit, so that the waters might not transgress His command, when He marked out the foundations of the earth, then I was beside Him, like a master workman ... (Proverbs 8: 22, 27-29)

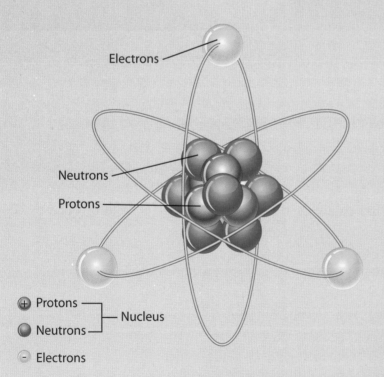

Electrons

Neutrons

Protons

⊕ Protons
● Neutrons — Nucleus
⊖ Electrons

Most problems that students have with mathematics go back to their early training where a poor foundation was laid. It is never too late to go back and strengthen your math.

Many properties of atoms have since been understood by studying their parts. Electrons have been shown to have a property that Benjamin Franklin called negative charge. This is a property of an electron or a description of its influence upon other objects. They attract protons that have positive charge. Objects with opposite charges (positive and negative) attract each other. Objects of the same charge repel each other. Electrons attract protons; electrons repel electrons; and protons repel protons.

If atoms have 8 protons, they are oxygen atoms, even though they may have different numbers of neutrons and electrons. An oxygen atom where the charges balance each other has 8 positive protons and 8 negative electrons. If an oxygen atom picks up 2 more electrons (which the oxygen you breathe does in metabolism), it will still have 8 protons (+8) but 10 electrons (-10). This gives the oxygen atom an overall charge of -2. With a -2 charge it can attract positively charged atoms and repel other negatively charged atoms. When the charges of an atom or molecule are not balanced, it will have an overall positive or negative charge and is called an **ion**. Just as the number of protons establishes the identity of an atom, the electrons are involved in the chemical bonds that hold atoms together. The neutrons (with no charge) contribute to the mass of atoms not their charges. Within the cells of our bodies, oxygen atoms snatch up electrons from food molecules that we have eaten and through a series of steps release most of the energy that keeps us alive.

Another important property of atoms is their mass. The mass of an atom is mainly a property of its protons and neutrons. The mass of an electron is negligible when contrasted to that of a proton or neutron. Remember that the metric unit of mass is the gram. A gram, however, is too large to measure the mass of something so small as a proton or neutron. That would be like trying to find the weight of an ant using tons. A much smaller unit of mass used is the **atomic mass unit (amu)**. A proton and a neutron each have a mass very close to 1 amu. An oxygen atom with 8 protons and 8 neutrons has a combined mass of 16 amu (8 + 8).

All hydrogen atoms have 1 proton and most of them have no neutron giving them an atomic mass of 1 amu. Some hydrogen atoms have 1 proton and 1 neutron and have an atomic mass of 2 amu. These atoms are called **deuterium**. Not many but some hydrogen atoms have 1 proton and 2 neutrons with an atomic mass of 3 amu and are called **tritium**. Notice that even though they have different atomic masses (called

The Three Isotopes of Hydrogen

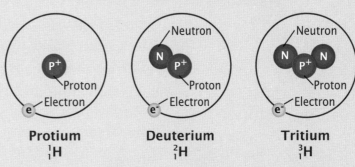

| Protium | Deuterium | Tritium |
| $^1_1$H | $^2_1$H | $^3_1$H |

their **atomic mass number,** which is the sum of the numbers of protons and neutrons), they still each only have 1 proton. This means that they are still hydrogen atoms. Atoms that have the same number of protons and different numbers of neutrons are **isotopes** of each other. The word isotope is a relational term. An atom by itself cannot be an isotope. It is an isotope when contrasted to another atom. It is like the term brother or sister. Most people think about radioactivity when they hear the word isotope, but most isotopes are not radioactive.

Look at your **periodic table of the elements**. In the box in the upper left is the large letter H, which stands for hydrogen. The numbers in the box describe hydrogen. The number 1 in the upper right corner of the box is the atomic number which is the number of protons. The number in the lower right corner of the box is 1.008 (1.0079 on some periodic tables). This is the average atomic mass of all of the isotopes of hydrogen that exist in nature. Most hydrogen atoms have an atomic mass of 1 amu, so the average comes out pretty close to 1.

Examples of several elements and their atomic masses are as follows.

Element	Symbol	Atomic Mass	Rounded Off Atomic Mass
Hydrogen	H	1.00794	1
Carbon	C	12.0107	12
Nitrogen	N	14.0067	14
Oxygen	O	15.9994	16
Sodium	Na	22.98976	23
Sulfur	S	32.065	32
Chlorine	Cl	35.453	35.5
Potassium	K	39.0983	39
Calcium	Ca	40.078	40
Iron	Fe	55.845	56

We use the atomic masses rounded off quite often, if we do not need the increased accuracy, because they are easier to work with. Notice that some of the symbols do not appear to go with the name of the element – such as K for potassium. The symbols are based upon the older Latin names of the elements, some of which have been changed over the years (like *Plumbus* for lead).

When atoms combine to form molecules, their atomic masses are added together to give their **molecular masses**. An example is sodium chloride (NaCl). The molecular mass of NaCl is the sum of 1 Na and 1 Cl.

$$
\begin{array}{llll}
1\ \text{Na} & = & 1 \times 23.0 & = 23.0 \\
\underline{1\ \text{Cl}} & = & 1 \times 35.5 & = \underline{35.5} \\
1\ \text{NaCl} & = & & 58.5\ \text{amu}
\end{array}
$$

Each NaCl molecule has a molecular mass of 58.5 amu. It is very difficult to measure out 58.5 amu of NaCl. As when preparing solutions in lesson 4, the mole is used. We can say that 1 mole of NaCl molecules has a mass of 58.5 grams. This is much easier to deal with.

Another example is the molecular mass of $CaCO_3$ (calcium carbonate).

$$
\begin{array}{llll}
1\ Ca & = 1 \times 40\ g/mole & = 40 \\
1\ C & = 1 \times 12\ g/mole & = 12 \\
\underline{3\ O} & \underline{= 3 \times 16\ g/mole\quad = 48} \\
1\ CaCO_3 & = & 100\ g/mole
\end{array}
$$

How many moles are there in 14 grams of Na_2SO_3 (sodium sulfite)?

The molecular mass of Na_2SO_3 is ...

$$
\begin{array}{llll}
2\ Na & = 2 \times 23\ g/mole & = 46 \\
1\ S & = 1 \times 32\ g/mole & = 32 \\
\underline{3\ O} & \underline{= 3 \times 16\ g/mole\quad = 48} \\
1\ Na_2SO_3 & = & 126\ g/mole
\end{array}
$$

14 grams of Na_2SO_3 would be much less than a mole of Na_2SO_3.

$$14 \text{ grams} / 126 \text{ grams per mole} = 0.11 \text{ mole of } Na_2SO_3$$

The expression grams divided by grams/mole gives grams/grams, which is 1, and the mole comes up on the top of the fraction. It is similar to $1/(1/2) = 2$.

$$\frac{1}{\frac{1}{2}} = 1 \div \frac{1}{2} = 1 \times \frac{2}{1}$$

$$\frac{g}{g/mole} = \frac{gram}{\frac{gram}{mole}} = gram \div \frac{gram}{mole} = \cancel{gram} \times \frac{mole}{\cancel{gram}} = mole$$

If you are uncertain of the math used here, follow the procedure and later, when you cover this in a math course, you will understand it better. It is okay to learn a procedure and then understand why you did it that way later.

PERIODIC TABLE

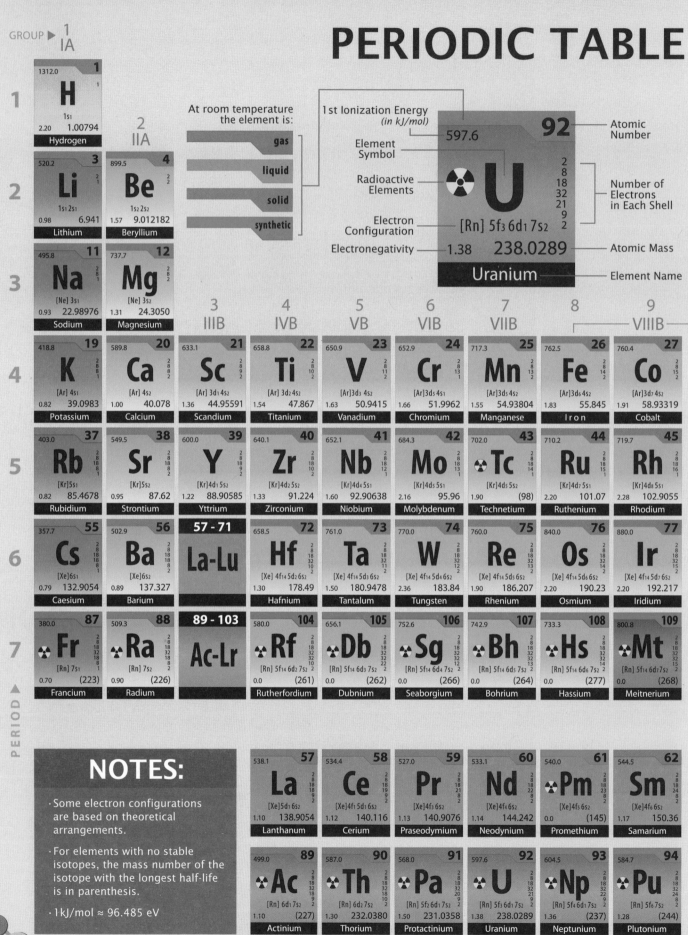

NOTES:

· Some electron configurations are based on theoretical arrangements.

· For elements with no stable isotopes, the mass number of the isotope with the longest half-life is in parenthesis.

· 1kJ/mol ≈ 96.485 eV

OF THE ELEMENTS

Subcategory in the metal-metalloid-nonmetal trend
(color of background)

- Alkaline Metal
- Lanthanide
- Transition Metal
- Alkaline Earth Metal
- Actinide
- Post-transition Metal
- Metalloid
- Polyatomic Nonmetal
- Diatomic Nonmetal
- Noble Gas
- Unknown Chemical Properties

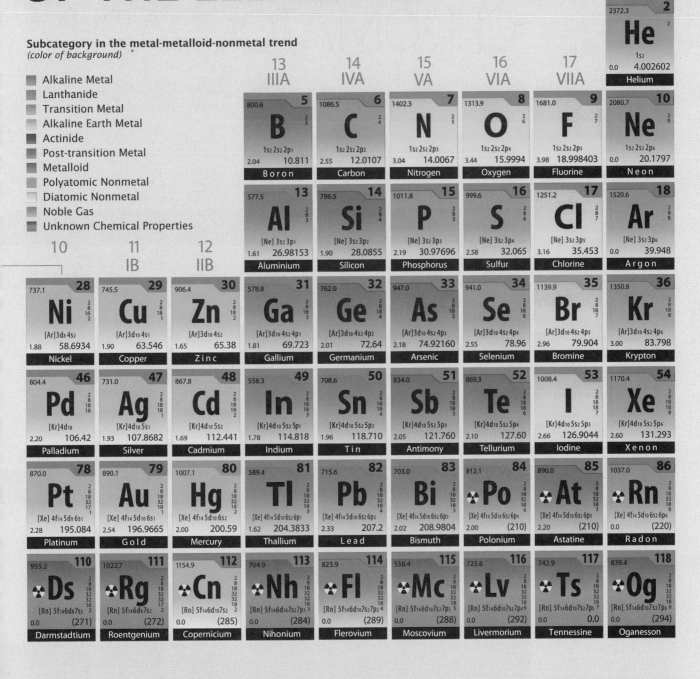

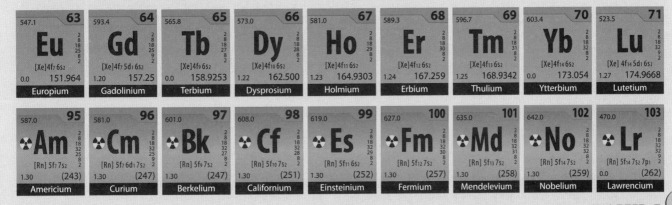

LABORATORY 5

CHROMATOGRAPHY

REQUIRED MATERIALS

- Filter paper

- Toothpicks

- Food coloring (yellow, blue, green)

- M&M's™, Kool Aid™, or other colorful substitutes (markers, paint, etc.)

- Weigh boats (or other small containers for mixing)

- Pencil

- Beaker (100 ml)

INTRODUCTION

Chromatography can come in many different forms and involve varying levels of complexity. An important principle in science is that you have to be careful about your goals. If you need more accuracy in your measurements, the equipment is expensive. A rule of thumb is that if you double your accuracy, you will probably increase your cost tenfold.

PURPOSE

Lesson 5 deals with the larger molecules made up of many atoms. Some molecules, such as DNA and proteins, are extremely large, consisting of thousands of atoms. This lab exercise will give you experience using an important tool in chemistry for separating and identifying molecules. Remember that even though many molecules are extremely large for molecules, they are still very small. When you use chromatography, you do it to separate different kinds of molecules from each other, identify them, and determine how much of each you have. In this exercise, we are trying to separate different molecules from each other.

PROCEDURE

1. Take a piece of circular filter paper and with a pencil trace out a circle about 2 cm in from the outside rim and one 4 cm in diameter around the center of the paper. A lid from a jar or rim of a glass may work well to make a nice even circle.

2. With a pencil mark the numbers 1–4 around the outer edge of the circle and poke a small hole right in the center of the filter paper with the pencil.

3. Using 4 toothpicks, place 2 drops of each of the following liquids at 4 points around the inner circle (of the filter paper with the hole in the middle) in line with the numbers. (For any of these you can make a colorful substitute.)

    A. Chartreuse food coloring (12 drops of yellow food coloring and 1 drop of green food coloring).

    B. Turquoise food coloring (5 drops of blue food coloring and 1 drop of green food coloring).

    C. Place of few drops of water on a few brown or tan M&Ms™. You can eat them afterward.

    D. Take some powdered Kool Aid™ and add a few drops of water to make a paste.

4. Take a second piece of filter paper and roll it into a tube that is narrow on one end and wide at the other end.

5. Make a drawing of the first piece of filter paper with the drops of liquid on it.

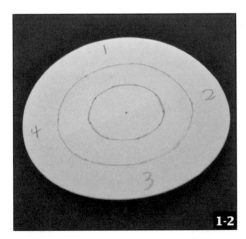

1-2

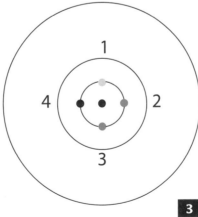

3

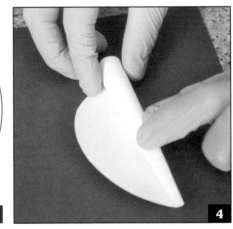

4

6. Pour water into a 100 ml beaker up to the 30 ml mark.

7. Place the narrow part of the filter paper cylinder into the hole of the first piece of filter paper so that the cylinder (the narrow end) hangs down below it.

8. Place the cylinder, while attached to the first filter paper, into the water in the beaker. Do it so that the first filter paper rests upon the rim of the beaker. The cylinder will act as a wick, drawing the water up to the first filter paper. When the water reaches the first filter paper it will spread out through the circle.

9. Let the water spread out toward the outer rim of the first filter paper until it reaches the outer circle that you drew. As the water spreads out, it should draw some of the colored dots with it. The different molecules in the drops should be separated from each other as they travel with the water at different rates. The water molecules are polar (having + and − charged ends) so they will draw polar molecules along with them. The larger molecules will not move as fast with the water as the smaller molecules. As well, the molecules that are more polar (having stronger + and − charges) will move faster with the water.

10. When the water reaches the outer circle that you drew on the filter paper, pull the filter papers out of the water to stop additional water from coming up the cylinder. Look at the color patterns formed on the top filter paper and draw what you see. The colors may fade as the water dries, so you will need to make a permanent record of your results. With a pencil, mark on the filter the areas of different color or, as an alternative, you can take a picture of the chromatogram.

If you desire, you can repeat the procedure with different substances of your choosing.

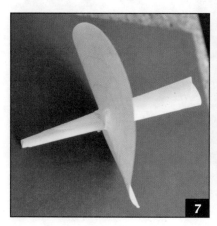

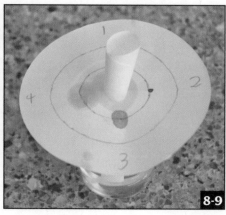

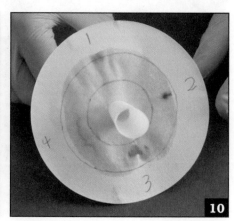

11. For your report, describe the procedure that you followed, describe the 4 samples that you used, and your results, including the drawing of the filter paper. Summarize your results, describing what happened to each of the 4 samples. As best you can, describe the numbers of different molecules in each sample and how they contrast with each other.

12. The water is pulled up the filter paper by what is called capillary action. Look up capillary action and in your own words write a complete sentence describing the process. Write it so that someone else without a background in chemistry will understand it. Do not use vocabulary that you do not understand. This is part of your report.

Chromatography means "writing in colors," because it usually produces a colored pattern where substances are separated from each other. There needs to be something to attract the substances (stationary phase) in a mixture and something to carry the substances (mobile phase) past the attracting material. As the substances are carried past the materials attracting them, they are separated from each other. This is because some of the substances are attracted more than others. The attracting material can be on a piece of paper or in a column. The carrying material can be a liquid or a gas.

Discoveries
in CHEMISTRY

Example of the separation of food coloring compounds by chromatography.

CHAPTER 6

PREPARING MOLAR SOLUTIONS

OBJECTIVES AND VOCABULARY

At the conclusion of this lesson the student should have an understanding (as evidenced by successfully completing the quiz at the end of this lesson) of:

1. How to determine how many moles of solute are needed to prepare a given volume of a solution of known molarity

2. How to determine how many grams of solute are needed to prepare a solution of known volume and molarity.

MOLAR
CONCENTRATION

MOLAR
SOLUTION

earning how to prepare a **molar solution** is to chemistry what learning how to use a paint brush is to an artist.

The first step is to determine how many moles of solute are needed to prepare a given volume of solution.

Study the following examples.

How many moles of NaCl do you need to prepare 1 liter of 3M NaCl? 3M means 3 moles of solute per liter of solution, so 1 liter of 3M NaCl needs 3 moles of NaCl to be added to 1 liter of water.

What if you only want to prepare 500 ml (1/2 of a liter) of 3M NaCl? For half of a liter, you need half as much NaCl or 3/2 or 1.5 moles of NaCl.

$$\frac{1}{2} \text{ liter} \times 3 \frac{\text{moles}}{\text{liter}} = \frac{3}{2} \text{ moles} = 1.5 \text{ moles}$$

What if you need 3 liters of 2M $C_{12}H_{22}O_{11}$ (sucrose, sugar)? Each liter of solution needs 2 moles of sucrose, so 3 liters needs 3 times as much or 3 x 2 = 6 moles of sucrose.

$$3 \text{ liter} \times 2 \frac{\text{moles}}{\text{liter}} = 6 \text{ moles}$$

If you need ¾ of a liter of 2M sucrose, how many moles of sucrose do you need? 1 liter of 2M sucrose needs 2 moles of sucrose, so ¾ of a liter of 2M sucrose needs ¾ x 2 or 6/4 or 1.5 moles of sucrose. This is properly written out as ...

$$\frac{3}{4} \text{ liter x } 2 \frac{\text{moles}}{\text{liter}} = 1.5 \text{ moles}$$

The units of liter cancel each other, leaving moles.

$$\frac{3}{4} \cancel{\text{liter}} \text{ x } 2 \frac{\text{moles}}{\cancel{\text{liter}}} = \frac{6}{4} \text{ moles} = 1.5 \text{ moles}$$

In practice, we cannot measure out moles of solute on a scale to make our solutions because the weight of a mole is different for every different solute. Usually we can find how many moles of solute we need but then we need to convert the moles to grams so that we can measure them out in the laboratory. The following examples demonstrate how to find how many grams of solute you need when you know how many moles you need.

Study the following examples.

If you want to make a 2M (2 molar or 2 moles/liter) solution of NaCl, how many grams of NaCl would you add to 1 liter of water?

You need 2 moles of NaCl to add to a liter of water and there are 58.5 grams of NaCl per mole (as determined earlier). 2 moles of NaCl are 2 moles x 58.5 grams/mole, which is 117 grams of NaCl in 2 moles. So, to make a 2M solution of NaCl, add 117 grams of NaCl to a liter of water.

1 Na	=	1 x 23.0 =	23.0
1 Cl	=	1 x 35.5 =	35.3
1 NaCl	=		58.5 g/moles

$$2 \cancel{\text{mole}} \text{ x } 58.5 \frac{\text{grams}}{\cancel{\text{mole}}} = 117 \text{ grams}$$

How would you make a 1.5 M $CaCO_3$ solution?

There are 100 grams in a mole of $CaCO_3$, so there is (1.5 x 100) 150 grams of $CaCO_3$ in 1.5 moles of $CaCO_3$. Add 150 grams of $CaCO_3$ to a liter of water to make the 1.5 M solution.

1 Ca	=	1 x 40	=	40
1 C	=	1 x 12	=	12
3 O	=	3 x 16	=	48
1 $CaCO_3$	=			100 g/mole

$$1.5 \cancel{\text{mole}} \text{ x } 100 \frac{\text{grams}}{\cancel{\text{mole}}} = 150 \text{ grams}$$

The molecular mass of $KMnO_4$ is (1 x K + 1 x Mn + 4 x O, which is 1 x 39 + 1 x 55 + 4 x 16, which is 39 + 55 + 64 = 158 grams/mole) 158 grams/mole. One half of a mole of $KMnO_4$ is ½ x 158 grams/mole, which is 79 grams. To make a 0.5 M $KMnO_4$ solution add 79 grams of $KMnO_4$ to a liter of water.

$$
\begin{array}{llll}
1\text{ K} & = & 1 \text{ x } 39 = & 39 \\
1\text{ Mn} & = & 1 \text{ x } 55 = & 55 \\
4\text{ O} & = & 4 \text{ x } 16 = & 64 \\
\hline
1\text{ } KMnO_4 = & & & 158 \text{ g/mole}
\end{array}
$$

$$\frac{1}{2} \text{ mole x } 158 \frac{\text{grams}}{\text{mole}} = 79 \text{ grams}$$

The molecular mass of NaCl is 58.5 grams/mole (23 + 35.5) and ½ of a mole of NaCl is 58.5/2 = 29.25 grams of NaCl. To make a liter of .5M NaCl, add 29.25 grams to a liter of water. To make 500 ml of 0.5 M NaCl, add half as much to make half as much solution. 29.25 grams/2 = 14.63 grams of NaCl to 500 ml of water, which is ½ of a liter.

$$
\begin{array}{llll}
1\text{ Na} & = & 1 \text{ x } 23.0 = & 23.0 \\
1\text{ Cl} & = & 1 \text{ x } 35.5 = & 35.3 \\
\hline
1\text{ NaCl} & = & & 58.5 \text{ g/moles}
\end{array}
$$

$$\frac{1}{2} \text{ mole x } 58.5 \frac{\text{grams}}{\text{mole}} = 29.25 \text{ grams}$$

$$\frac{1}{2} \text{ liter x } 29.25 \frac{\text{grams}}{\text{liter}} = 14.63 \text{ grams}$$

2 M $C_{12}H_{24}O_{11}$ would be 2 moles of $C_{12}H_{24}O_{11}$ dissolved into a liter of water. To make 500 ml, which is ½ of a liter, divide the 2 moles in half to give 1 mole. The molecular mass of $C_{12}H_{24}O_{11}$ is (12 x C + 24 x H + 11 x O = 12 x 12 + 24 x 1 + 11 x 16 = 144 + 24 + 176 = 344) 344 g/mole. To dissolve 1 mole of $C_{12}H_{24}O_{11}$ in 500 ml of water, dissolve 344 grams of $C_{12}H_{24}O_{11}$ in 500 ml of water.

$$
\begin{array}{llll}
12\text{ C} & = & 12 \text{ x } 12 = & 144 \\
24\text{ H} & = & 24 \text{ x } 1 = & 24 \\
11\text{ O} & = & 11 \text{ x } 16 = & 176 \\
\hline
1\text{ } C_{12}H_{24}O_{11} = & & & 344 \text{ g/mole}
\end{array}
$$

$$2 \text{ moles x } 344 \frac{\text{grams}}{\text{mole}} = 688 \text{ grams}$$

$$\frac{1}{2} \text{ liter x } 688 \frac{\text{grams}}{\text{liter}} = 344 \text{ grams}$$

LABORATORY 6

PREPARATION OF MOLAR SOLUTIONS

REQUIRED MATERIALS

- Table salt

- Baking soda

- Weigh boats

- Scale

- Beaker (100 ml)

- Beaker (250 ml)

- Laboratory scoop

- Stirring rod

PURPOSE

In this lab exercise, you will gain experience preparing solutions of given **molar concentrations.** This is an important skill used often in chemistry. Even if you are not the one preparing the solutions, this exercise will enable you to know how to work with the solutions when you know their molar concentrations.

PROCEDURE

Part 1: Prepare 100 ml of a 1 M (1 molar) NaCl solution.

The molecular mass of NaCl is (23.0 + 35.5) 58.5 grams/mole.

1 Na	=	1 x 23.0 =	23.0
1 Cl	=	1 x 35.5 =	35.5
1 NaCl	=		58.5 g/moles

So, if you prepared a liter of 1 M NaCl, you would add 58.5 grams of NaCl to a liter of distilled water. In this exercise, you are to prepare 100 ml (milliliters) of 1 M NaCl — 100 ml is 1/10 of a liter, so you need to use 58.5 grams/10 or 5.85 grams. If your scale measures grams to 1/10 of a gram, round the value of 5.85 grams off to 5.9 grams.

$$\frac{1}{10} \text{ liter x } 58.5 \frac{\text{grams}}{\text{liter}} = 5.85 \text{ grams}$$

1. Measure 100 ml of distilled water into a 250 ml beaker.

2. Measure 5.85 grams of NaCl.

3. Add the NaCl to your 100 ml sample of water and stir it until it completely dissolves.

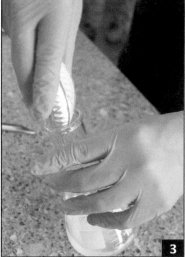

Part 2: Prepare 50 ml of a 3 M NaHCO$_3$ (sodium bicarbonate, baking soda) solution.

The molecular mass of NaHCO$_3$ is (23 + 1 + 12 + [3 x 16]) 84 grams/mole.

Na	=	23	=	23
H	=	1	=	1
C	=	12	=	12
3 O	=	3 x 16	=	48
NaHCO$_3$	=			84 g/mole

If you were preparing a liter of 3 M NaHCO$_3$, you would add 3 x 84 or 252 grams to a liter of water.

$$3 \text{ moles x } 84 \frac{\text{grams}}{\text{mole}} = 252 \text{ grams}$$

But you are to prepare 50 ml of 3 M NaHCO$_3$. 50/1,000 = 0.05, which means that 50 ml is 0.05 liter. If you would add 252 grams of NaHCO$_3$ to a liter of water, you will add 0.05 x 252 or 12.6 grams of NaHCO$_3$ to 50 ml of water.

$$0.05 \text{ liter x } 252 \frac{\text{grams}}{\text{liter}} = 12.6 \text{ grams}$$

1. Measure 50 ml of distilled water into a 100 ml beaker.

2. Measure 12.6 g of NaHCO$_3$.

3. Add the NaHCO$_3$ to your 50 ml sample of water and stir it until it completely dissolves.

In your report, show how you would prepare 500 ml of a 1.5 M solution of C$_{12}$H$_{22}$O$_{11}$ (sucrose, table sugar).

MOLAR RELATIONSHIPS

Comparison of molar concentrations and volume:

LAB INSIGHTS

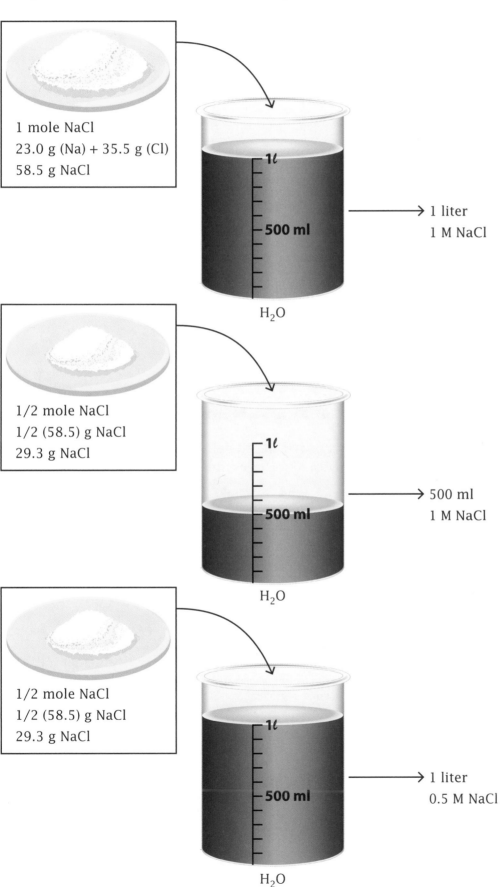

1 mole NaCl
23.0 g (Na) + 35.5 g (Cl)
58.5 g NaCl

1ℓ

500 ml

H₂O

→ 1 liter
1 M NaCl

1/2 mole NaCl
1/2 (58.5) g NaCl
29.3 g NaCl

1ℓ

500 ml

H₂O

→ 500 ml
1 M NaCl

1/2 mole NaCl
1/2 (58.5) g NaCl
29.3 g NaCl

1ℓ

500 ml

H₂O

→ 1 liter
0.5 M NaCl

CHAPTER 7

CHEMICAL REACTIONS

OBJECTIVES AND VOCABULARY

At the conclusion of this lesson the student should have an understanding (as evidenced by successfully completing the quiz at the end of this lesson) of:

1. Knowing what is meant by the terms chemical reaction, spontaneous reactions, kinetics, catalysts, enzymes, and metabolism

2. How to distinguish between endergonic reactions, exergonic reactions, precipitation reactions, acid-base reactions, and oxidation-reduction reactions.

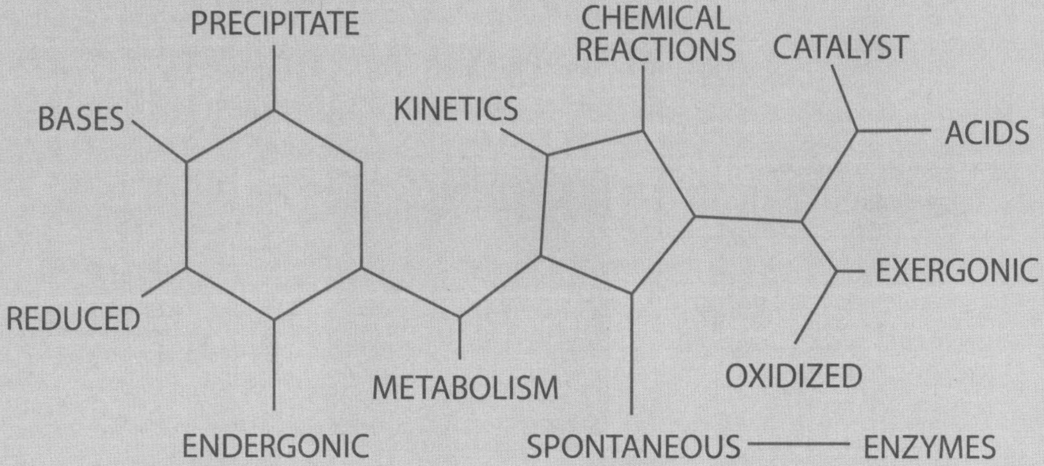

PRECIPITATE   CHEMICAL REACTIONS   CATALYST

BASES   KINETICS   ACIDS

REDUCED   EXERGONIC

METABOLISM   OXIDIZED

ENDERGONIC   SPONTANEOUS —— ENZYMES

Chemical reactions involve molecules coming apart with their atoms being put back together in diverse arrangements, making new molecules. In the cells of our bodies, molecules that we have eaten are taken apart and the atoms are rearranged into molecules that we need. Someone once said that he would not eat chicken because he knew what chickens ate. This could be a problem if you ate the stomach of the chicken right after the chicken ate a big juicy worm. In reality, the worm gets broken apart by small stones in the gizzard of the chicken and passed into the chicken's stomach. Larger molecules of proteins, carbohydrates, nucleic acids, and lipids are taken apart into smaller molecules that are absorbed into the chicken's blood stream. Within the cells of the chicken, these molecules are further taken apart and rearranged into other molecules.

For many chemical reactions, it is easy to tell if a reaction has occurred, but some are more difficult to recognize.

Chemical reactions occur when reactant molecules collide with each other in specific ways at appropriate speeds. For example, it is very difficult to knock the front bumper off a car by hitting it in the back with a bicycle going 5 miles an hour. Molecules are objects that have to

Water freezes at 0˚ C (32˚ F) and boils at 100˚ C (212˚ F).

hit each other for them to come apart. The motion of molecules is caused by their kinetic energy (energy of motion) that depends upon temperature. At higher temperatures they move faster and chemical reactions occur in less time (meaning that they come apart and reform faster) and at lower temperatures they move slower and chemical reactions take longer. As a general rule, chemical reactions go twice as fast with every 10˚ C (about 21˚ F) rise in temperature. ˚C (Celsius) is the metric unit of temperature. To convert ˚F to ˚C, subtract 32 from the ˚F and divide the result by 1.8.

$$\frac{(F - 32)}{1.8} = C$$

For example, if it is 72˚ F, subtract 32 and get 40. Then divide 40 by 1.8 and get 22.2˚ C. Degrees Celsius is a handy unit because water freezes at 0˚ C and boils at 100˚ C at sea level.

Chemists use the word **spontaneous** to mean that a chemical reaction can occur without saying how long it will take. For example, a piece of paper will react with oxygen in the air and burn into a pile of ashes. Even though, without adding heat to the paper, it could take several years, a chemist would still say that it was spontaneous. You have probably noticed that a very old piece of paper gradually turns yellow and brown. This is the result of the slow burning process. It is being oxidized, meaning that oxygen is removing electrons from the atoms in the paper, which takes apart chemical bonds between the atoms leaving ashes.

Chemists use the word spontaneous *to mean that a chemical reaction can occur without saying how long it will take.*

Remember that electrons participate in the bonds that hold atoms together. The branch of chemistry that deals with how long reactions take is **kinetics**.

Sometimes we need to make a reaction go faster. This can be done by raising the temperature, which causes the molecules to move faster, colliding with each other more often. It is not always desirable to change the temperature — such as in your body. In this case, a **catalyst** is used. These can be metal atoms or large protein molecules that attract the reacting (reactant) molecules to each other causing them to interact sooner. In our cells chemical reactions occur at a faster pace to keep us alive with large protein catalysts called **enzymes**. These reactions (called **metabolism**) would take much longer otherwise.

Just having the right molecules and energy is not enough to support or create life. The enzymes would have had to be already in place at the beginning of creation for metabolism to occur. Besides that, the enzymes are often embedded in cell membranes for organization. This way the products of one reaction become the reactants for the next reaction. This is evidence of God's remarkable intelligence and design at the beginning and continuation of life.

When molecules are rearranged into new molecules, energy is required. New bonds between atoms in new combinations may be stronger (store more energy) than the atoms of the molecules before a reaction. If this occurs with molecules in a container, the container gets colder because the reaction is taking energy from its surroundings. This is what happens when you have a cold pack for a sports injury. When you squeeze the cold pack, you mix the ingredients and they react with each other, taking energy from their surroundings. Reactions that absorb energy are called **endergonic**. The beginning of the word endergonic (en-) is similar to the word "in," so energy goes into the products. For other reactions where the bonds of the product molecules have less energy, the reaction gives off energy. The reactants have more energy and the products have less energy, so energy is released as the products are formed. These reactions are called **exergonic**. The word exergonic begins like the word exit, so energy goes out. When you squeeze a heat pack for an injury, you mix the reactants in the pack and it gives heat to the surroundings.

Precipitation Reactions

These reactions produce a product that settles out of the solution. An example is where a lead nitrate solution is mixed with a potassium iodide solution. The lead nitrate $Pb(NO_3)_2$ and potassium iodide KI solutions are clear. They look just like water. When mixed, a thick yellow material settles out to the bottom of the container. The word precipitation means to come out of solution.

$$Pb(NO_3)_2 \text{ (aq)} + 2 KI \text{ (aq)} \rightarrow PbI_2 \text{ (s)} + 2 KNO_3 \text{ (aq)}$$

In this equation, the (aq) means aqueous or dissolved in water and (s) means solid — not in solution. The PbI_2 is a **precipitate,** meaning that it is not in solution. Your clue that it is a precipitation is that there is a solid in the products that was not there before.

The chemical reaction (called precipitation) when potassium iodide is mixed with lead nitrate.

Acid-Base Reactions

Chemical reactions are also described by the kinds of molecules involved. Many reactions in water involve **acids** and **bases**. About 1 out of 10,000,000 (10 million) H_2O molecules comes apart, becoming H^+ and OH^-. The H^+ is an acid and the OH^- is a base. They are found wherever water is present. A common acid base reaction is where hydrochloric acid HCl (found in stomach acid and in most toilet bowl cleaners) is mixed with sodium hydroxide NaOH (also called lye, which is used as a caustic oven cleaner).

$$HCl + NaOH \rightarrow H_2O + NaCl$$

HCl comes apart as H^+ (acid) and Cl^- and NaOH comes apart as Na^+ +

acid + base → salt + water

OH^- (base). The acid H^+ combines with the base OH^- to form water H_2O. Acid-base reactions can be recognized by their products of H_2O and a salt such as NaCl.

Oxidation-Reduction Reactions

Many reactions involve the movement of electrons from some atoms to others. These are called oxidation-reduction reactions. When an atom loses an electron, it is **oxidized** (named after oxygen, which is a great "electron thief"). When an atom gains an electron, it is **reduced**. Oxidation and reduction reactions have to go together because an atom can only gain an electron when another atom loses it.

An example of an oxidation-reduction reaction is ...

$$Ca \text{ (s)} + 2 H^+ \text{ (aq)} \rightarrow Ca^{++} \text{ (aq)} + H_2 \text{ (g)}$$

Solid Ca (calcium) has as many protons (+ charge) as electrons (- charge). As a product Ca^{++} has 2 more positive charged protons than negative charged electrons. The Ca was oxidized, meaning that it lost electrons.

$$Ca \rightarrow Ca^{++} + 2 e^-$$

The H^+ hydrogen ions have 1 positive charged proton and no negative charged electrons. The H atoms in H_2 are not shown to have positive or negative charges, meaning that each hydrogen atom has an electron for every proton. The

When an atom loses an electron it is oxidized. When an atom gains an electron it is reduced.

negative electron balances the positive charge of each H atom. Each hydrogen atom is said to be reduced because it received an electron.

The calcium reduced the hydrogen and the hydrogen oxidized the calcium.

$$2\ H^+ + 2e^- \rightarrow H_2$$

Notice that the solid Ca is a reactant, not a product. In the precipitation reaction the solid was a product.

In later chapters, you will balance chemical reactions. A chemical equation has the reactants on the left and the products on the right. A balanced equation shows how many atoms are involved and how they are arranged in molecules. For example, the equation for the reaction that occurs in plants during photosynthesis in which plant cells take in carbon dioxide and water to produce glucose (a 6-carbon sugar) and oxygen molecules is ...

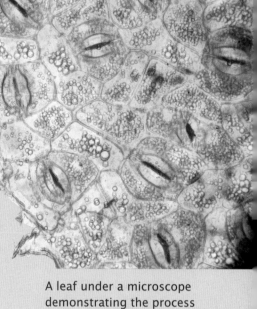

A leaf under a microscope demonstrating the process of photosynthesis.

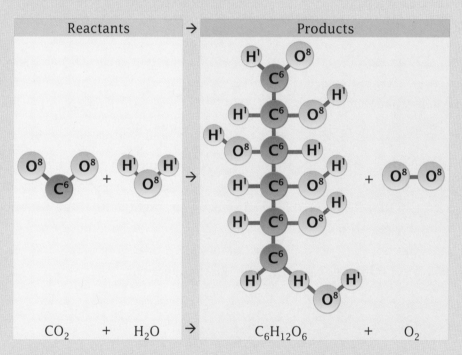

	Reactant	Product
C	1	6
O	2 + 1 = 3	6 + 2 = 8
H	2	12

$$CO_2 + H_2O \rightarrow C_6H_{12}O_6 + O_2$$

This shows which molecules are involved in the reaction but not how many of each molecule is involved. The equation still has to be balanced. The left side (reactants) has a total of 3 oxygen atoms (2 in CO_2 and 1 in H_2O) and the right side has 8 oxygen atoms (6 in $C_6H_{12}O_6$ and 2 in O_2). You cannot create 8 oxygen atoms from 3 oxygen atoms. Only God can do that. We have to show the same number of oxygen atoms on both sides of the equation. That applies to all of the other atoms (C and H) as well. Look at the same equation after it is balanced.

	Reactant	Product
C	1(**6**) = 6	6
O	2(**6**) + 1(**6**) = 18	6 + 2(**6**) = 18
H	2(**6**) = 12	12

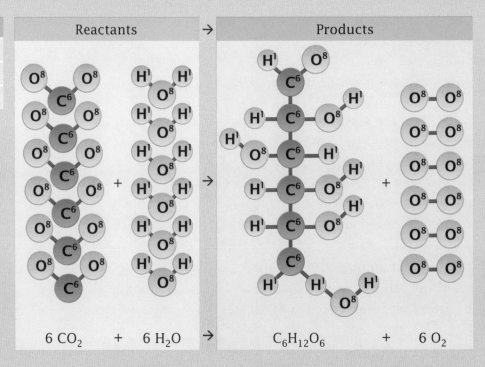

$$6\ CO_2\quad +\quad 6\ H_2O\quad \rightarrow\quad C_6H_{12}O_6\quad +\quad 6\ O_2$$

The 6 CO_2 molecules have 12 oxygen atoms because each CO_2 has 2 and there are 6 CO_2 molecules (6 x 2 = 12). Likewise, the 6 H_2O molecules have 6 oxygen atoms because each H_2O has 1 oxygen atom. This gives 12 plus 6 or 18 oxygen atoms on the left side of the equation.

On the right side of the equation, $C_6H_{12}O_6$ has 6 oxygen atoms and the 6 O_2 molecules have 12 oxygen atoms because each O_2 has 2 oxygen atoms. This gives 6 plus 12 or 18 oxygen atoms on the right side of the equation. Now the oxygen atoms balance, giving 18 oxygen atoms on both sides of the equation. This is vital since chemists are unable to create oxygen atoms from nothing nor to simply make them vanish.

Henry Louis Le Chatelier (1850–1936), after many observations, stated that a reaction at equilibrium when placed under stress that tends to upset the equilibrium becomes altered in the direction to relieve the stress. The reaction goes both ways.

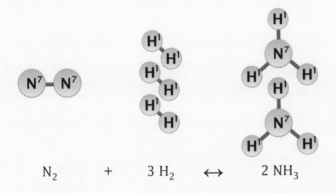

$$N_2 \quad + \quad 3\,H_2 \quad \leftrightarrow \quad 2\,NH_3$$

Saying that it is at equilibrium means that even though the reaction continues to go both ways, the total amounts of reactants and products remains the same. The N_2, H_2, and NH_3 are all gases, so if the pressure within the container where the reaction is occurring is increased, it will react to decrease the pressure. That means that more product (NH_3) will form and the amounts of N_2 and H_2 will decrease. There are 4 molecules of reactants for every 2 molecules of product, so the total number of gas molecules will decrease and the pressure will come back down. Notice that while all of this is going on, the equation remains balanced. In a later chapter, you will study chemical equilibrium and equilibrium constants, which is an extension of Le Chatelier's Principle.

If you add additional N_2 to the container, the gas pressure will increase and some of it will react with H_2 to form more NH_3, which will bring the pressure back down.

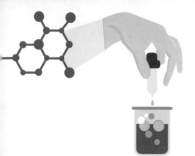

EVIDENCE OF A CHEMICAL REACTION

REQUIRED MATERIALS

- Small candle
- Empty can (able to hold the candle entirely)
- Vinegar
- Baking soda
- Long lighter (gas grill lighter)
- Small glass jar with disposable lid (will be damaged by experiment)
- Masking or duct tape
- Freezer-safe container large enough for the small glass jar
- Laboratory scoop

PURPOSE

How do you know that a chemical reaction has occurred? If water boils and evaporates, it changes state (liquid to gas) but it is still water, so no chemical reaction took place. In this exercise, you will determine if a new product was formed. It may be confusing if you are looking to see if the reactants are no longer present because a reaction can occur where not all of the reactants are changed into products.

PROCEDURE

1. Drip some wax from a burning candle into the inside floor of an empty can (soup, vegetable, etc.).

2. Blow out the candle and place the candle upright inside the can, supported by the melted wax.

   As the wax cools, it secures the candle.

3. Pour some vinegar (acetic acid) into the can, about half an inch deep around the base of the candle.

4. Carefully relight the candle. (Do not singe the hair around your knuckles.)

5. Carefully place about half of a teaspoon of sodium bicarbonate (NaHCO$_3$) into the vinegar, using the laboratory scoop, and observe what happens when the vinegar and sodium bicarbonate are mixed and what happens to the flame.

From your observations, did a chemical reaction occur? Was a new product formed when the vinegar and sodium bicarbonate were mixed? If a new product was formed, what did it do to the flame? Would the same thing happen to the flame if the sodium bicarbonate were not added to the vinegar? Try it. If a new product was formed, what do you think it could be? In your report, describe what you observed and your answers to these questions.

1. Take a small, expendable jar.

2. Fill it with water as full as you can get it and tightly secure the lid on the jar.

3. Take masking or duct tape and tape up the outside of the jar so that you can no longer see the glass.

4. Place it in a can, such as a coffee can, and place it in the freezer overnight.

5. The next day, remove it and describe your observations. Did a chemical reaction occur in the freezer? Why or why not? Remember that ice is still water, not a new substance.

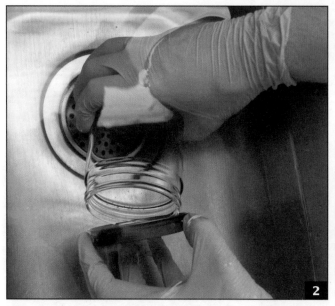

CHEMICAL VS PHYSICAL REACTIONS

In a chemical reaction, one chemical substance is transformed into another. In a physical reaction, the form of a substance changes but not the chemical composition.

Which of the following is a chemical reaction and which is a physical reaction?

LAB INSIGHTS

CHAPTER 8

CHEMICAL EQUATIONS I

OBJECTIVES AND VOCABULARY

At the conclusion of this lesson the student should have an understanding of balancing chemical equations.

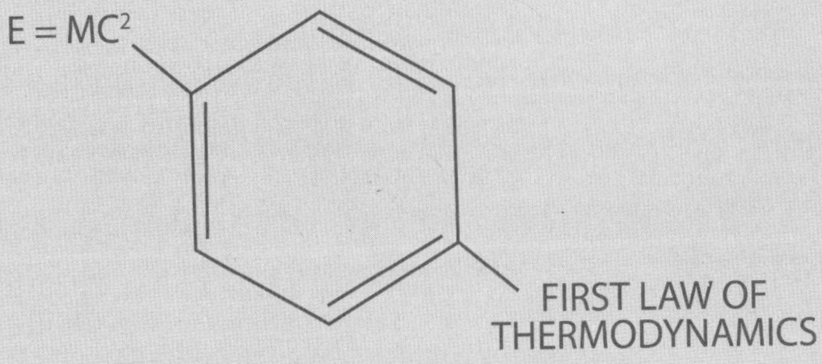

$E = MC^2$

FIRST LAW OF THERMODYNAMICS

An important principle to remember when balancing a chemical equation is that the atoms are conserved, meaning that they are neither created nor destroyed. The **First Law of Thermodynamics** says that energy cannot be created nor destroyed; only God can create or destroy energy (which is another strong argument saying God had to start the universe). Einstein's **E=mc²** says that atoms can be converted into pure energy ... but that would occur only in a nuclear reaction, like an atom bomb! This is why we have to come up with the same number of each atom on both sides of the equation.

The chemical formula for each molecule has to be left alone because you are not changing the identity of the molecules. For example, you cannot change Al_2O_3 to Al_2O_2 to balance an equation. That would be like trying to change a house cat into a lion. It just does not work that way.

Most chemical equations can be balanced by observation and matching up the numbers of each kind of atom on both sides of the equation.

For example …

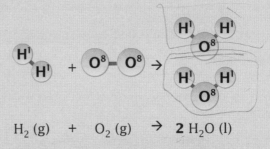

$$H_2 \text{ (g)} + O_2 \text{ (g)} \rightarrow H_2O \text{ (l)}$$

(g) means gas and (l) means liquid

	Reactant	Product
H	2	2
O	2	1

As written you have 2 H and 2 O on the left and 2 H and 1 O on the right, so the equation is not balanced. One O atom is lost and unaccounted for.

If you make the H_2O 2 H_2O instead, you have 2 O atoms on each side, giving you …

	Reactant	Product
H	2	2(2) = 4
O	2	1(2) = 2

$$H_2 \text{ (g)} + O_2 \text{ (g)} \rightarrow 2 \, H_2O \text{ (l)}$$

But now there are 2 H atoms on the left and 4 H atoms on the right. To give 4 H atoms on the left and the right, make H_2 2 H_2 giving …

	Reactant	Product
H	2(2) = 4	2(2) = 4
O	2	1(2) = 2

$$2 \, H_2 \text{ (g)} + O_2 \text{ (g)} \rightarrow 2 \, H_2O \text{ (l)}$$

This now gives 4 H atoms on each side and 2 O atoms on each side and the equation is balanced.

	Reactant	Product
H	2(4) = 8	2(4) = 8
O	2(2) = 4	1(4) = 4

If you came up with $4 \, H_2 \text{ (g)} + 2 \, O_2 \text{ (g)} \rightarrow 4 \, H_2O \text{ (l)}$, it would still be balanced. But to write the equation properly and make it more useful, use the smaller coefficients, giving $2 \, H_2 \text{ (g)} + O_2 \text{ (g)} \rightarrow 2 \, H_2O$. To make the equation useful, you need to know the ratios of the numbers of molecules, as you will see as we go along.

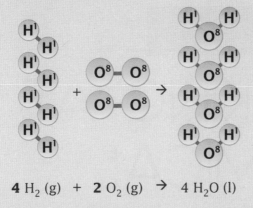

$$4 \, H_2 \, (g) \; + \; 2 \, O_2 \, (g) \; \rightarrow \; 4 \, H_2O \, (l)$$

If you had written H_2 (g) + ½ O_2 (g) → H_2O (l), it would not be correct because you cannot have half of a molecule.

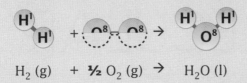

$$H_2 \, (g) \quad + \quad \tfrac{1}{2} \, O_2 \, (g) \; \rightarrow \quad H_2O \, (l)$$

	Reactant	Product
H	2	2
O	2(½) = 1	1

Consider this unbalanced equation.

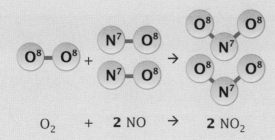

$$O_2 \quad + \quad NO \quad \rightarrow \quad NO_2$$

	Reactant	Product
O	2 + 1 = 3	2
N	1	1

Notice that there is 1 N (nitrogen) atom on each side of the equation. But there are 3 oxygen atoms on the reactant side and 2 on the product side. If you add 2 oxygen atoms to the product side and 1 to the reactant side, you will have 4 on each side.

	Reactant	Product
O	2 + 1(2) = 4	2(2) = 4
N	1(2) = 2	1(2) = 2

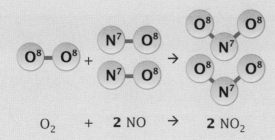

$$O_2 \quad + \quad 2 \, NO \quad \rightarrow \quad 2 \, NO_2$$

LABORATORY 8

LOOKING AT CHEMICAL REACTIONS WITH MOLECULAR MODELS

REQUIRED MATERIALS

• Molecular model kit

PURPOSE

In this exercise, the student will use molecular models to follow the steps of a series of chemical reactions.

PROCEDURE

Familiarize yourself with the different colored balls in the kit, which stand for different elements. Notice that the balls for hydrogen have 1 hole because hydrogen forms 1 covalent bond. The balls for oxygen have 2 holes because oxygen atoms form 2 covalent bonds. The balls for carbon have 4 holes for 4 covalent bonds.

Acetic Acid & Sodium Bicarbonate

Using the colored balls and pegs that connect the balls to each other, construct molecular models for acetic acid and sodium bicarbonate. Notice that the holes are positioned to give the covalent bonds proper directions between the atoms.

Double check to be sure that your models match the structure of the molecules as drawn below.

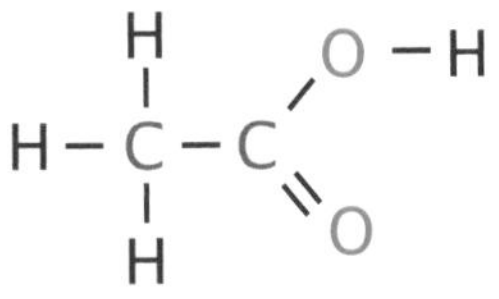

acetic acid

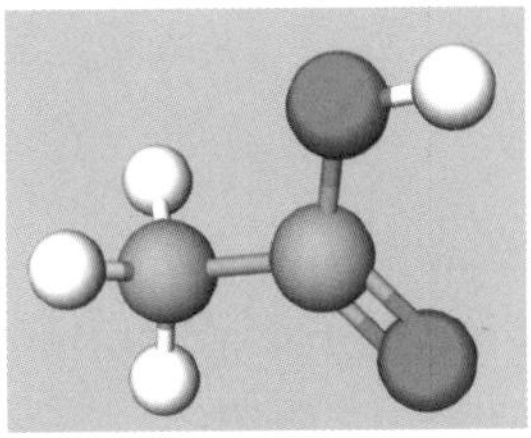

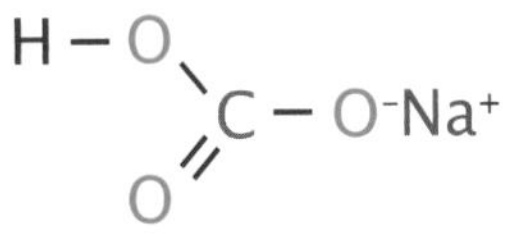

sodium bicarbonate

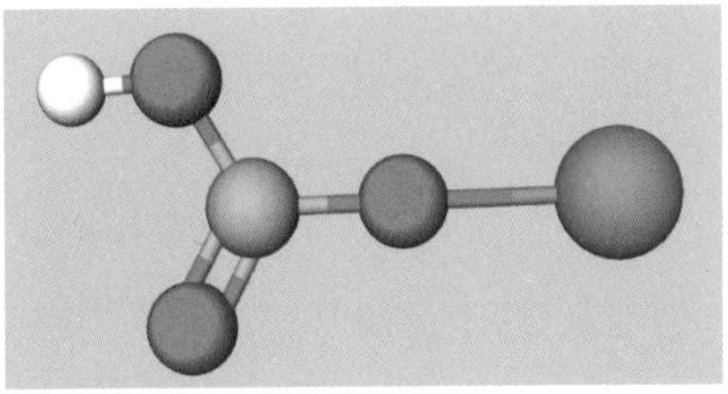

Here is the reaction between acetic acid and sodium bicarbonate.

Take the reactant molecules (acetic acid and sodium bicarbonate) apart enough so that you can form the products (sodium acetate, carbon dioxide, and water).

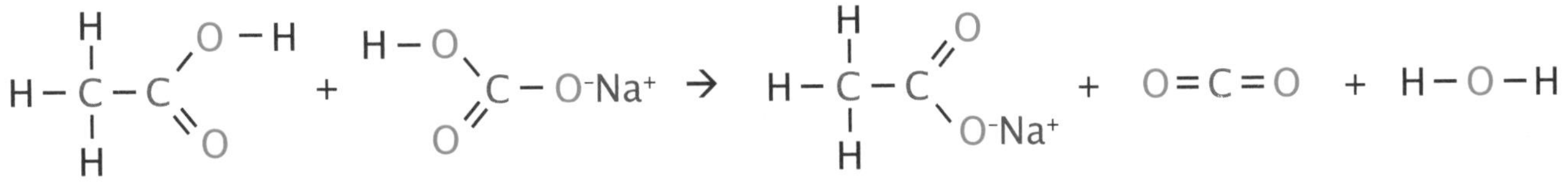

acetic acid   sodium bicarbonate   sodium acetate CO_2 H_2O

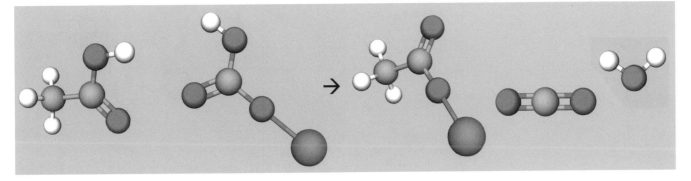

Hydrochloric Acid & Sodium Hydroxide

Make the models for HCl and NaOH. In the molecular model kit, the Cl is a halogen (green ball).

The HCl is an acid (hydrochloric acid) that reacts with the base NaOH (sodium hydroxide) to form water and sodium chloride.

$$HCl + NaOH \rightarrow H_2O + NaCl$$

Take the models for HCl and NaOH apart and form the molecular models for the products. This is easier than the first one.

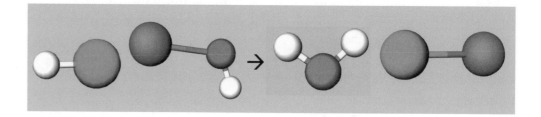

Glucose / Fructose

This one is a bit more difficult. Construct the glucose molecule (the 6 carbon sugar molecule that enters your metabolism) if you have enough and the right colored balls in your kit. Then convert it into the 6 carbon sugar molecule fructose (that is found in fruit). Notice that this would be a chemical reaction because glucose and fructose are different molecules. They both have the same elements and numbers of elements but differ in their arrangements. For this reason, they are called isomers of each other.

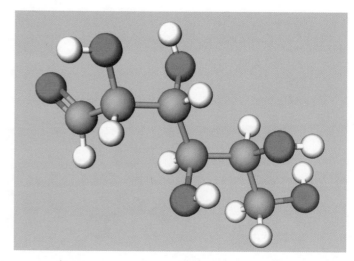

glucose

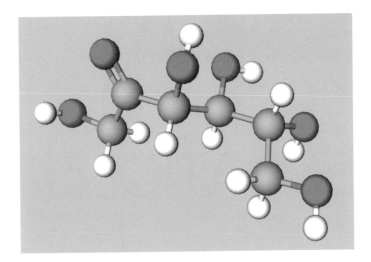

```
        H
        |
  H  —  C — O — H
        |
        C = O
        |
H — O — C — H
        |
  H  —  C — O — H
        |
  H  —  C — O — H
        |
  H  —  C — O — H
        |
        H
```

fructose

ISOMERS

LAB INSIGHTS

Is this an isomer of glucose and fructose? Hint — count the carbon atoms that are where the lines on the diagram come together.

Imagine the greater number of molecules that God created just by rearranging the atoms.

Each isomer is acted upon by a uniquely shaped enzyme that can recognize the shapes of the molecules. This enables each isomer to play a unique role in metabolism.

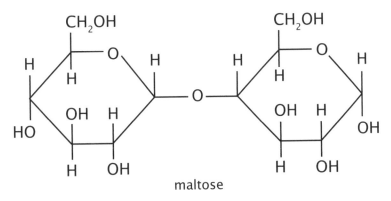

maltose

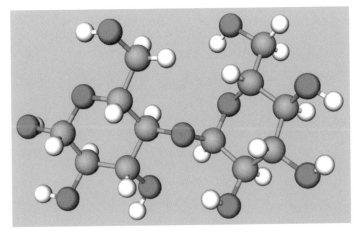

CHAPTER 9

CHEMICAL EQUATIONS II

OBJECTIVES AND VOCABULARY

The objective of this lesson is to gain further practice in balancing chemical equations.

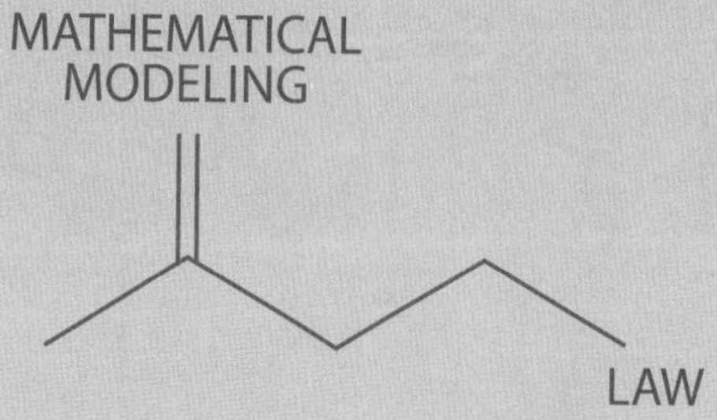

MATHEMATICAL
MODELING

LAW

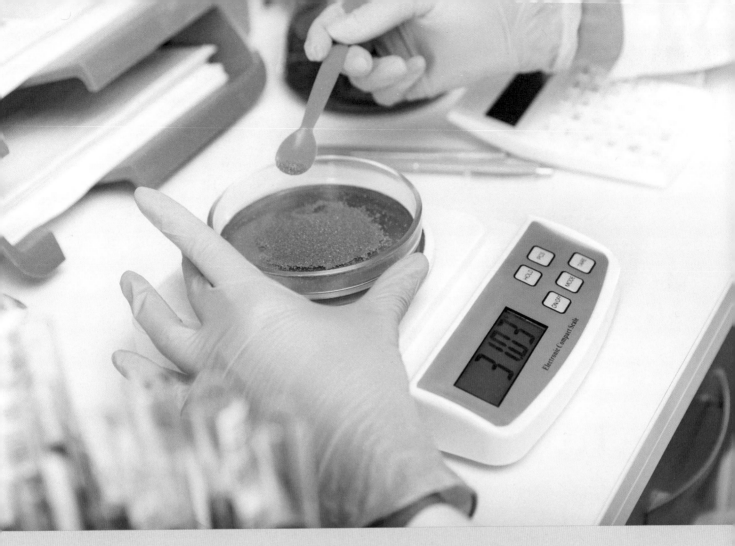

onsider the following unbalanced chemical equation, which would create a rather violent reaction you would never consider trying at home.

$$K\ (s) \quad + \quad H_2O\ (l) \quad \rightarrow \quad H_2\ (g) \quad + \quad KOH\ (aq)$$

There is 1 O atom on each side of the equation, so begin balancing with the H atoms. There are 2 H atoms in the H_2O of the reactants and 3 H in the H_2 and KOH of the products. Placing 2 KOH molecules on the product side gives a total of 4 H atoms on the product side. If we place 2 H_2O molecules on the reactant side, there will be 4 H atoms on each side of the equation and 2 O atoms on each side, giving ...

	Reactant	Product
K	1	1
H	2	2 + 1 = 3
O	1	1

	Reactant	Product
K	1	1(2) = 2
H	2(2) = 4	2 + 1(2) = 4
O	1(2) = 2	1(2) = 2

$$K\ (s) \quad + \quad \textbf{2}\ H_2O\ (l) \quad \rightarrow \quad H_2\ (g) \quad + \quad \textbf{2}\ KOH\ (aq)$$

	Reactant	Product
K	1(2) = 2	1(2) = 2
H	2(2) = 4	2 + 1(2) = 4
O	1(2) = 2	1(2) = 2

Now the H and O atoms are balanced, but there are 2 K atoms in the products and only 1 K atom in the reactants. If we place 2 K atoms on the reactant side, they are balanced. The final balanced equation is …

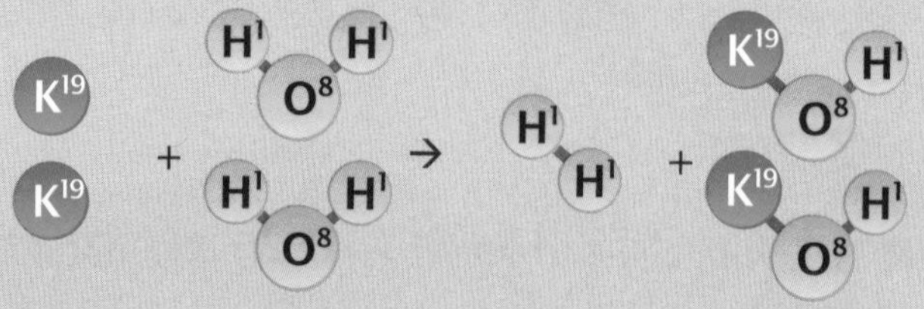

$$2\ K\ (s)\ +\ 2\ H_2O\ (l)\ \rightarrow\ H_2\ (g)\ +\ 2\ KOH\ (aq)$$

	Reactant	Product
N	1	1
H	3	2
O	2	1 + 1 = 2

The following unbalanced equation is not as easy to balance as it may look.

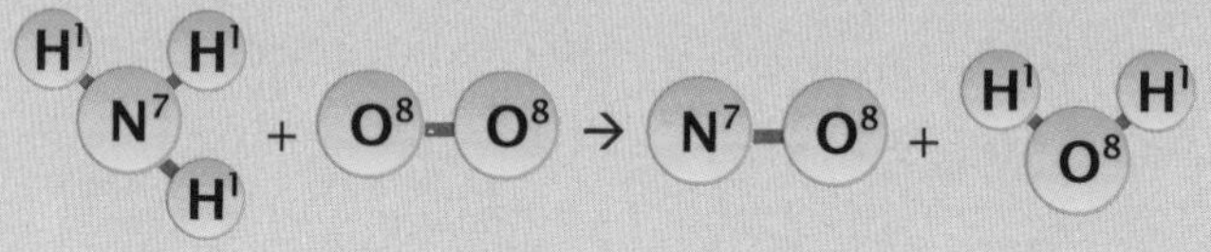

$$NH_3\ (g)\ +\ O_2\ (g)\ \rightarrow\ NO\ (g)\ +\ H_2O\ (g)$$

	Reactant	Product
N	1(2) = 2	1
H	3(2) = 6	2(3) = 6
O	2	1 + 1(3) = 4

Whenever there is an element by itself, like the O_2, it is usually best to balance it last because it can be changed without changing anything else. If we begin with the H atoms, there are 3 on the reactant side and 2 on the product side. 3 and 2 both go into 6, so let's try placing 6 of them on each side of the equation. To do this, we use $2\ NH_3$ and $3\ H_2O$, giving us …

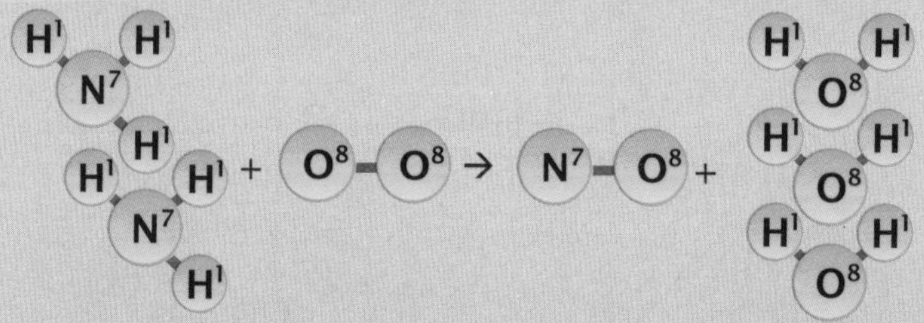

$$2\ NH_3\ (g)\ +\ O_2\ (g)\ \rightarrow\ NO\ (g)\ +\ 3\ H_2O\ (g)$$

	Reactant	Product
N	1(2) = 2	1(2) = 2
H	3(2) = 6	2(3) = 6
O	2	1(2) + 1(3) = 5

Now there are 6 H atoms on each side of the equation, and they are balanced. There are 2 N atoms on the reactant side, and only 1 N atom on the product side. If we place 2 NO molecules on the product side, there are now 2 N atoms on each side of the equation, giving …

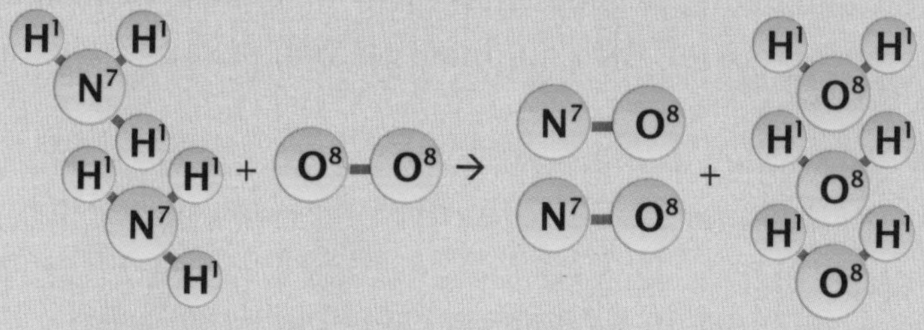

$$2\ NH_3\ (g)\ +\ O_2\ (g)\ \rightarrow\ 2\ NO\ (g)\ +\ 3\ H_2O\ (g)$$

There are now 2 O atoms on the reactant side and 5 O atoms on the product side. Even though this is awkward (temporarily), let's make the O_2 on the reactant side 2½ O_2. Now there are 5 O atoms on each side of the equation.

	Reactant	Product
N	1(2) = 2	1(2) = 2
H	3(2) = 6	2(3) = 6
O	2(**2.5**) = 5	1(2) + 1(3) = 5

$$2 \ NH_3 \ (g) \ + \mathbf{2 \ ½} \ O_2 \ (g) \rightarrow \ 2 \ NO \ (g) \ + \ 3 \ H_2O \ (g)$$

You cannot have ½ of an O_2 molecule, so we need to double everything in the equation, which gives us ...

	Reactant	Product
N	1(2)**(2)** = 4	1(2)**(2)** = 4
H	3(2)**(2)** = 12	2(3)**(2)** = 12
O	2(2.5)**(2)** = 10	1(2)**(2)** + 1(3)**(2)** = 10

$$4 \ NH_3 \ (g) \ + \ 5 \ O_2 \ (g) \ \rightarrow \ 4 \ NO \ (g) \ + \ 6 \ H_2O \ (g)$$

Here is another with a little different variation. In this case, aluminum is exposed to hydrochloric acid.

	Reactant	Product
Al	1	1
Cl	1	3
H	1	2

$$Al \ (s) \ + \ HCl \ (aq) \ \rightarrow \ AlCl_3 \ (aq) \ + \ H_2 \ (g)$$

The Al (aluminum) atoms are okay with 1 on each side. But there are 3 Cl (chloride) and 2 H on the product side and only 1 of each on the reactant side.

You can balance the Cl by ...

	Reactant	Product
Al	1	1
Cl	1**(3)** = 3	3
H	1**(3)** = 3	2

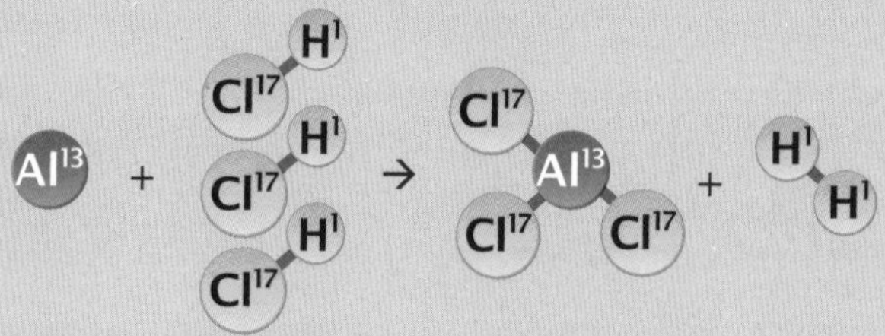

$$Al \quad + \quad \mathbf{3}\,HCl \quad \rightarrow \quad AlCl_3 \quad + \quad H_2$$

This makes the Cl happy, but now you have 3 H on the reactant side and 2 on the product side.

When you have 3 on one side and 2 on the other, many times it helps to multiply the 3 by 2 and the 2 by 3. This gives the common number of 6. It looks like this.

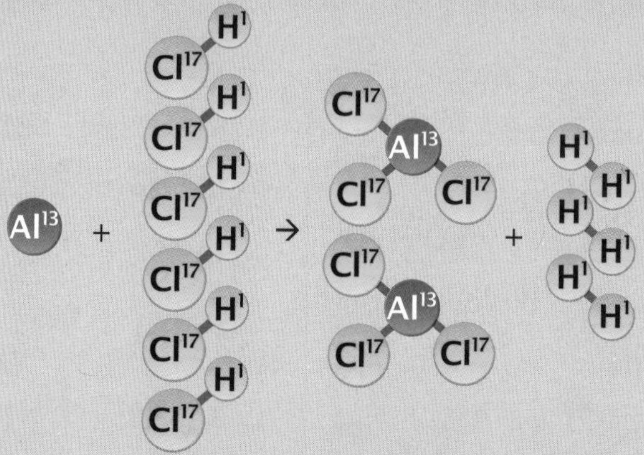

	Reactant	Product
Al	1	1**(2)** = 2
Cl	1(3)**(2)** = 6	3**(2)** = 6
H	1(3)**(2)** = 6	2**(3)** = 6

$$Al \quad + \quad \mathbf{6}\,HCl \quad \rightarrow \quad \mathbf{2}\,AlCl_3 \quad + \quad \mathbf{3}\,H_2$$

	Reactant	Product
Al	1**(2)** = 2	1(2) = 2
Cl	1(3)(2) = 6	3(2) = 6
H	1(3)(2) = 6	2(3) = 6

Now there are 6 Cl on each side and 6 H on each side. Now there is a problem with the Al. There are 2 on the product side and 1 on the reactant side. That is an easy fix. Because the Al is by itself on the reactant side you can just multiply it by 2 giving 2 Al on each side.

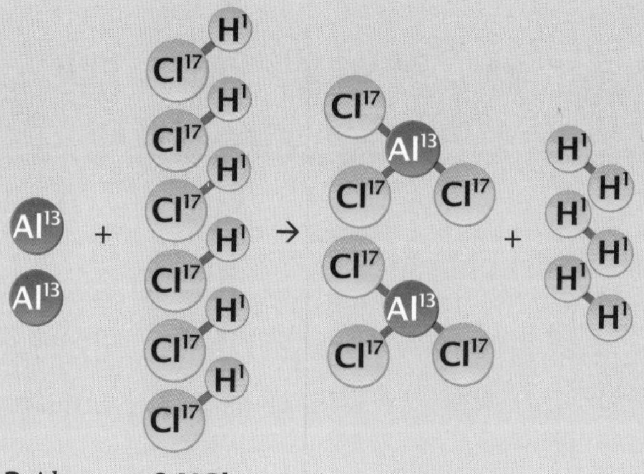

$$\mathbf{2}\,Al \quad + \quad 6\,HCl \quad \rightarrow \quad 2\,AlCl_3 \quad + \quad 3\,H_2$$

Consider this example that is a bit more complicated.

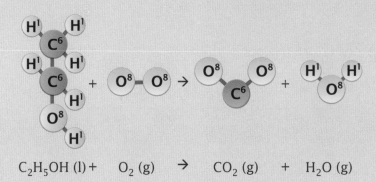

C_2H_5OH (l) + O_2 (g) → CO_2 (g) + H_2O (g)

	Reactant	Product
C	2	1
H	6	2
O	1 + 2 = 3	2 + 1 = 3

Here ethanol (ethyl alcohol) burns with oxygen to produce carbon dioxide and water. It is easiest to begin with the C (carbon) atoms because it is part of one and not both of the product molecules. The equation becomes ...

	Reactant	Product
C	2	1**(2)** = 2
H	6	2
O	1 + 2 = 3	2**(2)** + 1 = 5

C_2H_5OH (l) + O_2 (g) → **2** CO_2 (g) + H_2O (g)

This gives 2 C (carbon) atoms on each side. The H (hydrogen) is also in just one of the molecules of the products, so balance it next. C_2H_5OH (l) has 6 H atoms, so make it 3 H_2O (g) in the products. The equation becomes ...

	Reactant	Product
C	2	1(2) = 2
H	6	2**(3)** = 6
O	1 + 2 = 3	2(2) + 1**(3)** = 7

C_2H_5OH (l) + O_2 (g) → 2 CO_2 (g) + **3** H_2O (g)

This gives 2 C atoms on each side, 6 H atoms on each side, but 3 O atoms in the reactants and 7 O atoms in the products.

	Reactant	Product
C	2	1(2) = 2
H	6	2(3) = 6
O	1 + 2**(3)** = 7	2(2) + 1(3) = 7

If 3 O_2 (g) molecules are placed in the reactants, there are 7 O atoms on both sides of the equation. This gives ...

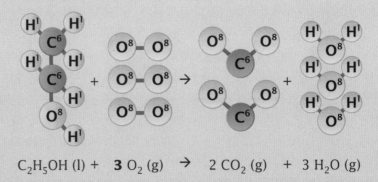

$$C_2H_5OH \text{ (l)} + \mathbf{3} \, O_2 \text{ (g)} \rightarrow 2 \, CO_2 \text{ (g)} + 3 \, H_2O \text{ (g)}$$

Now double check. This gives 2 C atoms on both sides of the equation, 6 H atoms on both sides, and 7 O atoms on both sides. This is now a balanced equation.

Discoveries *in* CHEMISTRY

The need to balance chemical equations is evidence of the order God created in the universe. As well, we have to balance the equations because we cannot create nor destroy matter or energy — only God can. Under special conditions (to be studied in physics), matter can be converted into energy, as shown by Albert Einstein's famous equation $E = mc^2$. In this equation, E is energy, m is the mass of matter converted into energy, and c is the speed of light (3.0×10^8 meters per second or 300,000,000 meters per second). When you square the speed of light, you get 9×10^{16} or 90,000,000,000,000,000. This is the energy released in a nuclear reaction. The creation is more complex than just matter and energy. There is a complex relationship that we do not yet understand very well between the two.

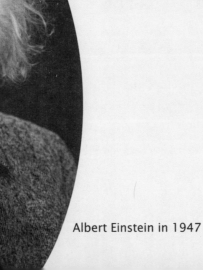

Albert Einstein in 1947

Satellite in orbit around the earth, as city lights illuminate the darkness below.

As in balanced equations and $E = mc2$, math represents the order that comes from the mind of God. If we can trust God with the complexities of the universe, we can certainly trust Him with the details of our lives.

These are like so many relationships in creation called **mathematical modeling.** From God's mind, there are so many levels of complexity of mathematical relationships around us. It is like recognizing the details that go into planning every step of flying a spacecraft to the moon and back. There are even mathematical relationships in predicting the impacts of food, water, hunting, competition, and predation upon deer populations. Our weather predictions are not limited by the lack of order but rather by our lack of understanding the order.

When you can assign accurate numbers to something, you can often see the order.

BALANCING CHEMICAL EQUATIONS WITH MOLECULAR MODELS

REQUIRED MATERIALS

- Molecular model kit

PURPOSE

This exercise is designed to enable the student to visualize the need to balance chemical equations. By constructing molecular models, the student can see how there has to be the same number of atoms of each element on both sides of a chemical equation.

PROCEDURE

Consider the reaction where methane is burned (oxidized) by oxygen.

$$CH_4 + O_2 \rightarrow CO_2 + H_2O$$

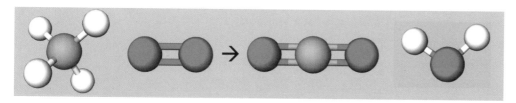

You can tell that this equation is not balanced. There are 3 oxygen atoms in the products and 2 oxygen atoms in the reactants. There are 4 hydrogen atoms in the reactants and only 2 in the products. Balance it before looking at the answer below.

If you place 2 H_2O molecules in the products, you have 4 hydrogen atoms on each side of the equation. The oxygen atoms are still unbalanced. With 2 O_2 molecules in the reactants, there are 4 oxygen atoms on each side of the equation. The equation becomes ...

$$CH_4 + 2\ O_2 \rightarrow CO_2 + 2\ H_2O$$

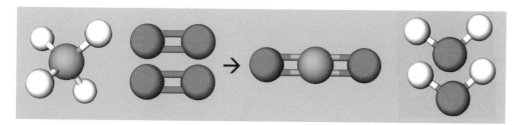

Using the molecular model set, make models for each of the reactant molecules. Make models for CH_4 and 1 O_2 molecule.

Take these models apart and make the models for the reactants CO_2 and H_2O.

Do you have enough atoms to make the models? Do you have any atoms left over from the reactants?

Go back and make another O_2 molecule for the reactants and use all of the atoms from the reactants to make the products. How many H_2O molecules did you make?

Look at the equation $O_2 + NO \rightarrow NO_2$. Is this equation balanced?

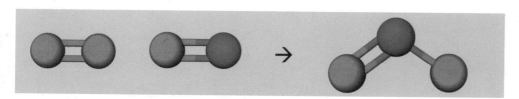

Discoveries in CHEMISTRY

Antoine Lavoisier

The Frenchman Antoine Lavoisier did much more than help develop the metric system. He provided evidence for and described the law of Conservation of Matter, which is another way of saying that matter and energy cannot be created nor destroyed except by God. This is the basis for balancing chemical equations. Whenever you see the word **law** in science, think of a restriction that God placed upon the creation from the very beginning.

Lavoisier also developed several measurement techniques in chemistry and wrote what is considered to have been the first modern chemistry book. The use of the word "modern" means that it represented a breakaway from alchemy. In the earlier days, chemistry was called alchemy, which was trying to turn metals of little value, such as lead, into gold. They were not concerned with laying a foundation of solid principles. A world that was masterfully designed should function by solid principles that were in the mind of the Designer.

The laboratory of Lavoisier in Musée des Arts et Métiers

Make a model for O_2 and one for NO. When you take the reactant models apart, how many of each element do you need to make NO_2 as products. Figure out how to balance the equation using the models. Write out the balanced equation.

Consider the equation H_2CO_3 (carbonic acid) + NaOH (sodium hydroxide) → Na_2CO_3 + H_2O. Construct a H_2CO_3 and a NaOH molecule using the molecular model kit. Use the orange balls (monovalent metal) for the Na.

$$H-O \diagdown \atop H-O \diagup \mkern-12mu C=O$$

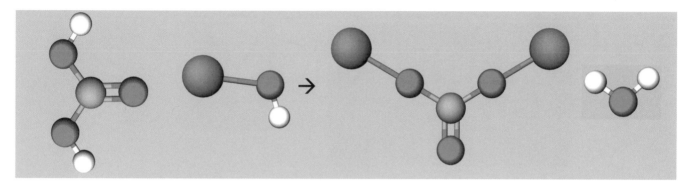

Take these molecular models apart and construct the products (Na_2CO_3 and H_2O) from those atoms. Do you have enough atoms to make the products? What atoms do you have to add to make the products? You have to use up all of the atoms from the reactants. Construct the product molecules, using up all of the reactant atoms and the atoms that you had to add. From this information write out the balanced equation for this reaction.

Your report consists of the drawings of the reactant and product molecules that you made and how many of each that you made and the balanced equation for H_2CO_3 + NaOH reaction.

CHAPTER 10

MOLES FROM CHEMICAL EQUATIONS

OBJECTIVES AND VOCABULARY

The objective of this lesson is to be able to find the number of moles of reactant and product needed in a chemical reaction.

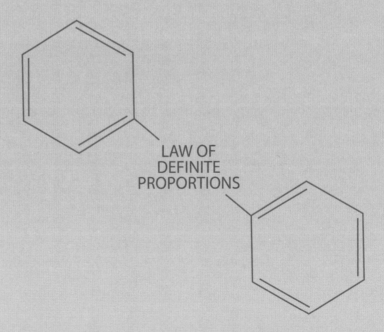

LAW OF
DEFINITE
PROPORTIONS

When you have a balanced equation, you can find out how much reactant you need and how much product you will get. In the examples below, you will see that the equation gives the number of moles, as well as the number of molecules. This is because a mole of molecules is always the same number of molecules. A mole is always the same number, just as a dozen is always 12.

In the reaction ...

$$C_2H_5OH \; (l) + 3 \; O_2 \; (g) \rightarrow 2 \; CO_2 \; (g) + 3 \; H_2O \; (g)$$

if there are 3 moles of C_2H_5OH to burn (react with O_2), how many moles of O_2 are needed?

$$\textbf{3} \; C_2H_5OH \; (l) + 3 \; O_2 \; (g) \rightarrow 2 \; CO_2 \; (g) + 3 \; H_2O \; (g)$$

From the equation, it can be seen that there are 3 moles of O_2 used for every mole of C_2H_5OH, so it will take 3 x 3 or 9 moles of O_2 to react with 3 moles of C_2H_5OH.

$$\textbf{3} \; C_2H_5OH + \textbf{3} \; \textbf{(}3 \; O_2\textbf{)} \rightarrow 2 \; CO_2 + 3 \; H_2O$$

If 3 moles of C_2H_5OH are used, how many moles of CO_2 and H_2O are produced? There are 2 moles of CO_2 for every mole of C_2H_5OH, so if 3 moles of C_2H_5OH are used, 2 x 3 or 6 moles of CO_2 are formed. There are 3 moles of H_2O produced for every mole of C_2H_5OH, so if 3 moles of C_2H_5OH are used, 3 x 3 or 9 moles of H_2O are produced.

$$3\ C_2H_5OH + \mathbf{3}\ (3\ O_2) \rightarrow \mathbf{3}\ (2\ CO_2) + \mathbf{3}\ (3\ H_2O)$$
$$3\ C_2H_5OH + 9\ O_2 \rightarrow 6\ CO_2 + 9\ H_2O$$

Consider this balanced reaction ...

$$3\ MnO_2\ (s)\ +\ 4\ Al\ (s)\ \rightarrow\ 3\ Mn\ (s)\ +\ 2\ Al_2O_3\ (s)$$

If 2 moles of Al (aluminum) are available for the reaction, how much MnO_2 will be used? The equation gives 4 moles of Al and 2 is half of 4, so half as much of the MnO_2 (manganese dioxide) will also be used, which is half of 3, which is 1½ moles of MnO_2.

$$\tfrac{1}{2}\ (3\ MnO_2) + \tfrac{1}{2}\ (4\ Al) \rightarrow 3\ Mn + 2\ Al_2O_3$$
$$1\tfrac{1}{2}\ MnO_2 + \mathbf{2}\ Al \rightarrow 3\ Mn + 2\ Al_2O_3$$

REACTIONS IN ACTION

Let's test your knowledge of fractions. If 1 mole of Al is available, how many moles of MnO_2 will be used? 1 mole of Al is ¼ of the 4 moles in the equation, so ¼ x 3 or ¾ mole of MnO_2 will be used. Once you see the pattern to these exercises, they will be easier to solve.

$$\tfrac{1}{4}\ (3\ MnO_2) + \tfrac{1}{4}\ (4\ Al) \rightarrow 3\ Mn + 2\ Al_2O_3$$
$$\tfrac{3}{4}\ MnO_2 + \mathbf{1}\ Al \rightarrow 3\ Mn + 2\ Al_2O_3$$

If 1 mole of Al is available, how much Al_2O_3 will be produced? Again 1 mole of Al, ¼ of the 4 moles in the equation, so for Al_2O_3, ¼ x 2 or ½ mole of Al_2O_3 will be produced.

$$\tfrac{1}{4}\ (3\ MnO_2) + \tfrac{1}{4}\ (4\ Al) \rightarrow 3\ Mn + \tfrac{1}{4}\ (2\ Al_2O_3)$$
$$\tfrac{3}{4}\ MnO_2 + \mathbf{1}\ Al \rightarrow 3\ Mn + \tfrac{1}{2}\ Al_2O_3$$

If 1 mole of Al is available, how much Mn will be produced? There are 3 moles of Mn produced for every 4 moles of Al used, so ¼ x 3 or ¾ mole of Mn will be produced.

$$¼ (3 MnO_2) + ¼ (4 Al) \rightarrow ¼ (3 Mn) + ¼ (2 Al_2O_3)$$
$$¾ MnO_2 + 1 Al \rightarrow ¾ Mn + ½ Al_2O_3$$

If 1 mole of Mn is desired from the reaction, how much Al is needed? For every 3 moles of Mn produced, 4 moles of Al are needed. 4/3 or 1 1/3 moles of Al are needed.

$$1/3 (3 MnO_2) + 1/3 (4 Al) \rightarrow 1/3 (3 Mn) + 1/3 (2 Al_2O_3)$$
$$1 MnO_2 + 4/3 Al \rightarrow 1 Mn + 2/3 Al_2O_3$$

Often you are called upon to first balance an equation and then find how many moles of product or reactant you need.

Balance this equation.

$$Al (s) + O_2 (g) \rightarrow Al_2O_3$$

	Reactant	Product
Al	1	2
O	2	3

Try it yourself first before looking at the following answer.

This is another situation with a 2 and 3 combination. Do the multiplying by 3 and 2. This gives …

Often you are called upon to first balance an equation and then find how many moles of product or reactant you need

$$Al + 3 O_2 \rightarrow 2 Al_2O_3$$

	Reactant	Product
Al	1	2(2) = 4
O	2(3) = 6	3(2) = 6

Now there are 6 oxygen atoms on each side of the equation. But now there is one Al in the reactants and 4 in the products. It is handy that the Al in the reactants is by itself. That way you can change it without changing anything else. Make it 4 Al in the reactants and the equation is balanced.

	Reactant	Product
Al	1(**4**) = 4	2(2) = 4
O	2(3) = 6	3(2) = 6

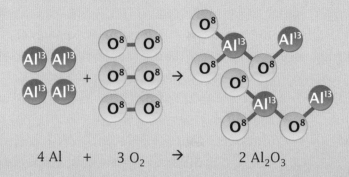

$$4\,\text{Al} \quad + \quad 3\,\text{O}_2 \quad \rightarrow \quad 2\,\text{Al}_2\text{O}_3$$

Now you can find how many moles of reactants are needed and how many moles of products can be made. If you want to get 2 moles of Al_2O_3, how many moles of O_2 do you have to use?

The balanced equation shows that 3 moles of O_2 are needed to make 2 moles of Al_2O_3, so the answer is 3 moles of O_2.

$$4\,\text{Al} + \mathbf{3}\,\text{O}_2 \rightarrow \mathbf{2}\,\text{Al}_2\text{O}_3$$

What if you only want to make 1 mole of Al_2O_3, how many moles of O_2 do you need? 1 mole is half of the 2 moles of Al_2O_3 in the equation. So, you need half as much of O_2, which is 3/2 or 1.5 moles of O_2.

$$\mathbf{1/2}\,(4\,\text{Al}) + \mathbf{1/2}\,(3\,\text{O}_2) \rightarrow \mathbf{1/2}\,(2\,\text{Al}_2\text{O}_3)$$
$$2\,\text{Al} + 1.5\,\text{O}_2 \rightarrow 1\,\text{Al}_2\text{O}_3$$

What if you want to make 3 moles of Al_2O_3, how many moles of O_2 do you need? 3 is 1.5 times 2, so you need 1.5 times 3 or 4.5 moles of O_2.

$$\mathbf{1.5}\,(4\,\text{Al}) + \mathbf{1.5}\,(3\,\text{O}_2) \rightarrow \mathbf{1.5}\,(2\,\text{Al}_2\text{O}_3)$$
$$6\,\text{Al} + 4.5\,\text{O}_2 \rightarrow 3\,\text{Al}_2\text{O}_3$$

When you are balancing an equation you have to end up with whole numbers (no fractions or decimals) because you cannot have a fraction of a molecule. The equation we used can mean 4 Al atoms and 3 O_2 molecules react to form 2 molecules of Al_2O_3. You can have a fraction of a mole just like you can have half of a dozen. So, it is okay for our calculations to use fractions of moles.

The French chemist Joseph Proust (1754–1826) developed techniques to measure the mass of the components of certain compounds. An example would be measuring the mass of hydrogen and oxygen in water molecules. He stated the **Law of Definite Proportions,** which states that the elements in a compound will always be in the same proportions no matter where they came from. You are demonstrating this when you make models of molecules with the colored balls and then rearrange them to make product molecules. The model for the water molecule will always have 2 balls for the hydrogen atoms and 1 ball for the oxygen atom. Everything that we have done up to this point depends upon this. The work of Proust gave further confidence in keeping the same atoms (even though they are rearranged) in the reactants and the products in a balanced equation.

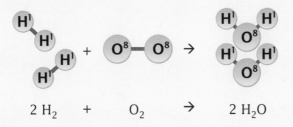

$$2\,H_2 \quad + \quad O_2 \quad \rightarrow \quad 2\,H_2O$$

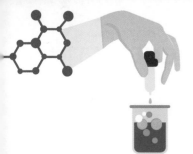

LABORATORY 10

ESTIMATING HOW MANY ATOMS ARE IN A MOLE

REQUIRED MATERIALS

- Erlenmeyer flask (250 ml)

- Graduated cylinder (10 ml)

- Vinegar

- Baking soda

- Balloon

- Scale

- Laboratory scoop/teaspoon measure

PURPOSE

In this exercise, the student is to visualize by measuring out a mole and using the molecular mass to determine how many atoms are in a mole. The number of atoms of carbon and oxygen is determined from the mass of a sample of CO_2.

PROCEDURE

1. Measure out 50 ml of vinegar (with the graduated cylinder) into the 250 ml Erlenmeyer flask.

2. Weigh a balloon on the scale.

3. Add a teaspoon of sodium bicarbonate to the flask and slip the balloon over the opening of the flask to trap the escaping CO_2 gas. Try to catch as much of the CO_2 as you can.

   The acetic acid (CH_3COOH) in the vinegar releases H+ ions that are the acid. The H+ ions react with the sodium bicarbonate to produce the CO_2.

   $$CH_3COOH + NaHCO_3 \rightarrow CH_3COONa + H_2O + CO_2$$

   The CO_2 is released as a gas that fills the balloon.

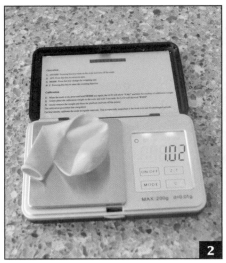

4. When it looks like the balloon is no longer expanding, quickly remove it and tie off the end to contain the CO_2.

5. Weigh the balloon with the CO_2.

6. Calculate the number of atoms of carbon and oxygen in the balloon.

   A. Subtract the weight of the balloon from the weight of the balloon and the CO_2 to get the mass of the CO_2 in the balloon.

$$1.34 \text{ g} - 1.02 \text{ g} = 0.32 \text{ g}$$

   B. Divide the mass of the CO_2 by the molecular mass of CO_2 (44 grams/mole, 12 g for C and 2 x 16 g for O) to get the number of moles of CO_2. This should be a very small number.

$$0.32 \text{ g} \div \frac{44 \text{ g}}{1 \text{ mole}} = 0.32 \text{ g} \times \frac{1 \text{ mole}}{44 \text{ g}} = 0.0073 \text{ mole}$$

   C. Multiply the number of moles of CO_2 by 6.022 and multiply the result by 10^{23} to get the number of molecules of CO_2. You are multiplying the moles of CO_2 by 6.022×10^{23} because there are 6.022×10^{23} items in a mole, just as there are 12 items in a dozen.

   The number of molecules of CO_2 equals the number of C atoms because there is 1 C atom in a CO_2 molecule.

$$0.0073 \times 6.022 \times 10^{23} = 4.396 \times 10^{21}$$

   D. Multiply the number of molecules of CO_2 by 2 to get the number of atoms of O because there are 2 atoms of O in every CO_2 molecule.

$$4,396,000,000,000,000,000,000 \times 2 = 8.792 \times 10^{21}$$

   This is the number of molecules of CO_2 formed and the number of C atoms and O atoms in the CO_2.

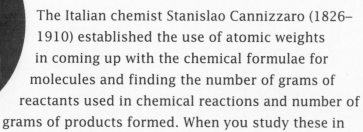

The Italian chemist Stanislao Cannizzaro (1826–1910) established the use of atomic weights in coming up with the chemical formulae for molecules and finding the number of grams of reactants used in chemical reactions and number of grams of products formed. When you study these in this text, it may seem like we have always known how to do it, but this was not the case.

Between the years of 1811 and 1858, it was a frustrating struggle to come up with the atomic weights of elements. Some felt that it would be impossible to know the atomic weights. In 1858, Cannizzaro determined that a molecule must have a whole number of atoms. He decided that the atomic weights had to be integral (whole number) multiples of some basic atomic weight — 1 for H (hydrogen) was used and 2 for H_2. He used Avogadro's principle that equal volumes of gas had an equal number of molecules to use the weights of equal volumes of gas compared to hydrogen to estimate their atomic weights.

He found that if he had the same volume of oxygen gas where the hydrogen has a mass of 2 grams, the oxygen had a mass of 32 grams. If the hydrogen has an atomic mass of 1 gram (H_2 is 2 grams), the oxygen would have an atomic mass of 16 grams.

H_2
2 grams

O_2
32 grams

CHAPTER 11

FINDING THE GRAMS OF REACTANT AND PRODUCT

OBJECTIVES AND VOCABULARY

The learning objective of this lesson is to find the number of grams of reactants and products.

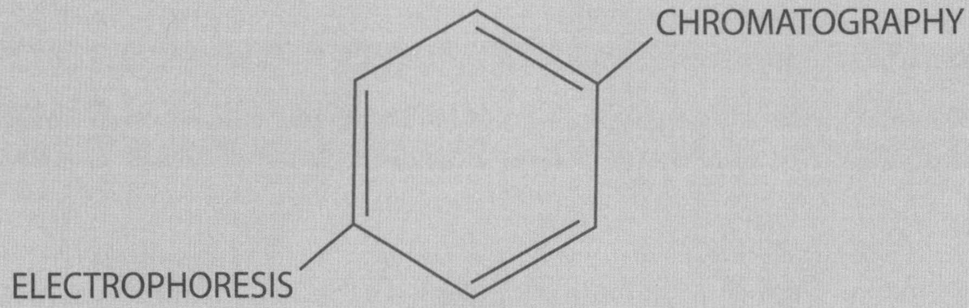

CHROMATOGRAPHY

ELECTROPHORESIS

Let's go back to an earlier equation.

$$2 \text{ K (s)} + 2 \text{ H}_2\text{O (l)} \rightarrow \text{H}_2 \text{ (g)} + 2 \text{ KOH (aq)}$$

How many grams of KOH would you get if you started out with 1 mole of K? In the equation, 2 moles of KOH are produced for every 2 moles of K used. So for every mole of K you get 1 mole of KOH. You may need to go back to chapters 2 and 3 to review molecular weights and page 48 for the atomic weights of some of the elements. We calculate the molecular weight of KOH as

K	=	39 g/mole
O	=	16 g/mole
H	=	1 g/mole
KOH	=	56 g/mole

If you started with 1 mole of K, you would get 1 mole of KOH, which is 56 grams.

If you started out with 10 grams of K, how many grams of KOH would you get? To find this, you first need to find out how many moles of K you have with the 10 grams. The balanced equation gives us the ratio of moles not grams. The atomic weight of K is 39 g/mole. The number of moles of K in 10 grams is ...

$$10 \text{ g} / (39 \text{ g/mole}) = 0.26 \text{ mole of K (10 divided by 39)}$$

$$\frac{10 \text{ gram}}{\dfrac{39 \text{ gram}}{\text{mole}}} = 10 \text{ gram} \div \frac{39 \text{ gram}}{\text{mole}} = 10 \text{ gram} \times \frac{\text{mole}}{39 \text{ gram}} = 0.26 \text{ mole}$$

This means that 0.26 mole of K will form 0.26 mole of KOH because K and KOH have the same number of moles on both sides of the equation. The number of grams of KOH is found by ...

$$0.26 \text{ mole} \times \frac{56 \text{ gram}}{\text{mole}} = 14.56 \text{ grams of KOH}$$

If you started out with 10 grams of K, how many grams of H_2 would be formed? Notice in the equation that there is 1 mole of H_2 formed for every 2 moles of KOH formed. So, if 0.26 mole of KOH is formed from 10 grams of K, 0.13 mole of H_2 (½ of 0.26) would be formed. This gives ...

$$0.13 \text{ mole} \times \frac{2 \text{ gram}}{\text{mole}} = 0.26 \text{ grams of } H_2 \text{ formed.}$$

Using the equation ...

$$4 \text{ NH}_3 \text{ (g)} + 5 \text{ O}_2 \text{ (g)} \rightarrow 4 \text{ NO (g)} + 6 \text{ H}_2\text{O (l)}$$

How many grams of NO are produced when 15 grams of NH_3 are used? First find out how many moles of NH_3 are used ...

N			=	14 g/mole
3 H	=	3 x 1	=	3 g/mole
NH_3			=	17 g/mole

$$(15 \text{ g of NH}_3) / (17 \text{ g/mole}) = 0.88 \text{ mole of NH}_3$$

$$\frac{15 \text{ gram NH}_3}{\dfrac{17 \text{ gram}}{\text{mole}}} = 15 \text{ gram} \div \frac{17 \text{ gram}}{\text{mole}} = 15 \text{ gram} \times \frac{\text{mole}}{17 \text{ gram}} = 0.88 \text{ mole}$$

In the equation, 4 moles of NO are produced when 4 moles of NH_3 are used, so the same number of moles of NO are produced as the number of moles of NH_3 used. If 0.88 mole of NH_3 are used, 0.88 mole of NO will be produced.

To find the number of grams of NO in 0.88 mole of NO, first find the molecular weight of NO.

$$
\begin{array}{lll}
\text{N} & = & 14 \text{ g/mole (from page 11)} \\
\text{O} & = & 16 \text{ g/mole} \\
\hline
\text{NO} & = & 30 \text{ g/mole}
\end{array}
$$

The number of grams in 0.88 mole of NO is ...

$$0.88 \text{ mole} \quad \text{x} \; \frac{30 \text{ gram}}{\text{mole}} = 26.4 \text{ grams of NO produced.}$$

Let's try a little more challenging example. How many grams of O_2 will it take to produce 25 grams of NO? First find out how many moles there are in 25 grams of NO.

$$\frac{25 \text{ gram}}{\dfrac{30 \text{ gram}}{\text{mole}}} = 25 \text{ gram} \div \frac{30 \text{ gram}}{\text{mole}} = 25 \text{ gram} \; \text{x} \; \frac{\text{mole}}{30 \text{ gram}} = 0.83 \text{ mole of NO}$$

In the equation, there are 4 moles of NO produced for every 5 moles of O_2 used. The moles of O_2 used to produce 0.83 mole of NO are ...

$$\frac{5}{4} \; \text{x} \; (0.83 \text{ mole}) = 1.04 \text{ moles of } O_2.$$

There are 2 atoms of O in every O_2 molecule, so the molecular weight of O_2 is ...

$$2 \; \text{x} \; \frac{16 \text{ gram}}{\text{mole}} = \frac{32 \text{ grams}}{\text{mole}}$$

The number of grams of O_2 needed is ...

$$1.04 \text{ mole} \quad \text{x} \; \frac{32 \text{ gram}}{\text{mole}} = 33.28 \text{ grams of } O_2.$$

LABORATORY 11

CHROMATOGRAPHY USING DIFFERENT SOLVENTS

REQUIRED MATERIALS

- Filter paper

- Beaker (100 ml)

- Scissors

- Pencil

- Pen / marker

- Paper clip and string, stirring rod and tape, or toothpick

- Toothpick (optional)

- Tape (optional)

- Isopropyl alcohol

- Olive oil

- Other color and liquid solvent

PURPOSE

This exercise demonstrates the effects of using solvents of different polarities to separate compounds in a mixture, known as **chromatography.**

PROCEDURE

1. Take 4 pieces of filter paper and cut them into long strips 1 inch wide.

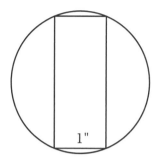

2. Place a dot of ink (any color) 1 inch up from the bottom of the strip in the middle. Mark the paper on each side of the dot with a pencil so that you can tell later where the dot was. Prepare 2 more pieces of filter paper the same way.

3. Test with water:

   A. Fill a 100 ml beaker with water about 1/2 inch deep.

   B. Suspend the filter paper in the beaker with the portion below the dot in the water. (You may tape the filter paper to the glass stirring rod, paper clip it to a piece of string stretched across the beaker, or push a toothpick through the filter paper to accomplish this). The paper must be suspended vertically in the middle of the beaker without touching the sides.

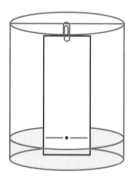

   C. Allow the water to travel up the filter paper (by capillary action) until it gets almost to the top. Do not let it reach the top.

   D. Remove the filter paper and lay it on a piece of paper towel.

   E. Mark with a pencil where the different colors have traveled up the paper.

      Water is a very polar molecule (with a + and a – charged end), so it attracts other polar molecules. The more polar molecules in the ink should travel farther with the water and the non-polar molecules should stay at the bottom. Molecules that are not as polar as water

should go part of the way with the water. Describe with complete sentences what you observe on the filter paper. What does this say about the compounds in the ink? Attach the dried filter paper to your report.

4. Repeat step 3 with a beaker of rubbing alcohol (usually 70 percent isopropyl alcohol) instead of water.

   Isopropyl alcohol is not as polar as water, so it should pull those compounds that are not as polar as water. Afterward, mark the filter paper as before and write out the description of your results. As before, attach the dried filter paper to your report.

5. Repeat step 3 with a beaker of olive oil instead of water or isopropyl alcohol.

   Olive oil is non-polar, so the results should be quite different again. Report your results as before.

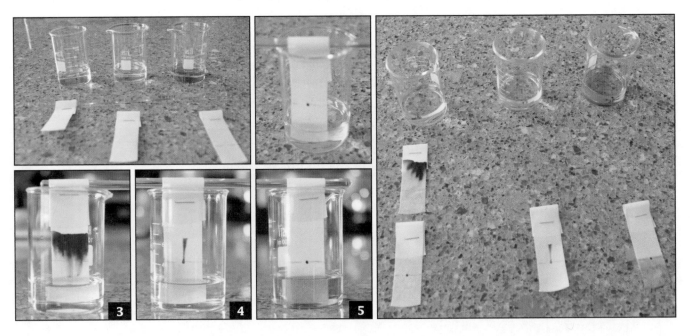

6. Prepare the fourth strip of filter paper using something of your choosing besides ink. Perform step 3 a fourth time using whichever solvent you desire. Before you do it, write out as part of your procedure what you predict the results will be. Do the procedure and evaluate your prediction. In this case, you do not suffer a prophet's judgment if your prediction does not come true. Base your prediction on what you observed with the first 3 procedures, using the polarity of the substance and the polarity of the solvent.

ELECTROPHORESIS

Electrophoresis is a technique of separating proteins in an electric field where one end of the apparatus is positive charged and the other end is negative charged. The proteins have positive and negative charges that cause them to be attracted to one end of the apparatus or the other. Positive-charged proteins are attracted to the negative end of the apparatus and the negative-charged proteins are attracted to the positive end of the apparatus. Smaller proteins move faster and further, and larger proteins move slower and a shorter distance. This is similar to chromatography, except the proteins that are separated are not colored. The proteins are pulled through material called agar (similar to Jell-O®) or across a moist surface.

LAB INSIGHTS

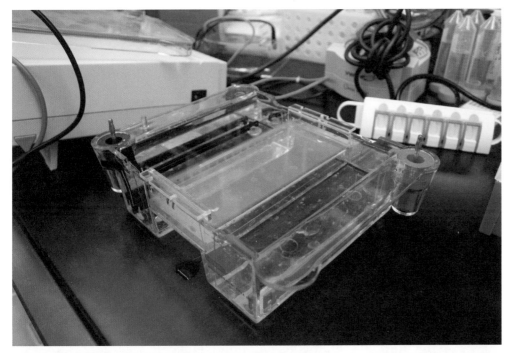

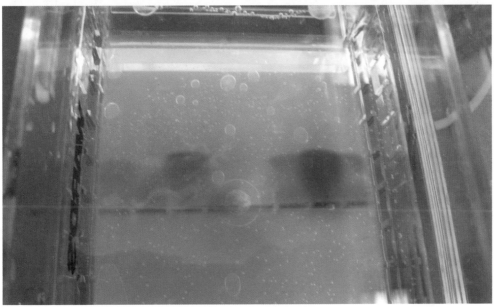

The process of electrophoresis showing the movement of charged particles in the gel, under the influence of the electric field.

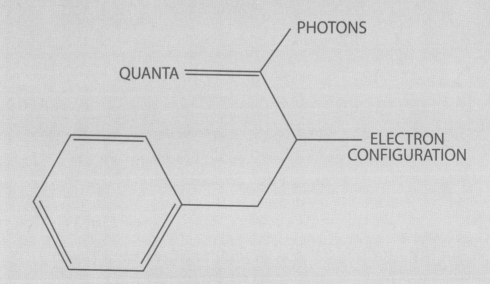

CHAPTER 12

ELECTRON CONFIGURATIONS

OBJECTIVES AND VOCABULARY

At the conclusion of this lesson the student should have an understanding (as evidenced by successfully completing the quiz at the end of this lesson) of:

1. Electron orbitals

2. Diagraming the electron configurations of electron orbitals

3. The distinction between metals and non-metals.

QUANTA

PHOTONS

ELECTRON
CONFIGURATION

The last four lessons dealt with a very important tool in chemistry — the chemical equation. The following lessons deal with another very important tool in chemistry — the periodic table. Some people memorize it but do not know what to do with it and some only memorize parts that they need and know how to use it. It does not need to be memorized any more than a phone book does. You need to be able to identify its major parts and know how to use it.

One of the major goals of science is to predict what happens under certain conditions. An example is weather forecasting.

> *One of the major goals of science is to predict what happens under certain conditions*

Sometimes it may seem to be less accurate, but it is much better than it used to be. The major goal of this lesson is to gain a basic understanding of the energy levels of electrons in atoms so that you can predict the outcomes of chemical reactions. Instead of memorizing hundreds of chemical reactions, it is far more profitable to learn a few basic principles that you can use to predict their outcomes. Could you imagine what it

Chemical reactions involve the breaking of chemical bonds between atoms and making new bonds in different combinations.

would be like if we had no idea as to what would happen when different things were mixed together? This could make for some rather interesting meals.

Chemical reactions involve the breaking of chemical bonds between atoms and making new bonds in different combinations. The number of protons in the nucleus of an atom determines its identity as an element but the electrons outside of the nucleus determine the chemical bonds that an atom can make with other atoms. Remember in an earlier lesson how oxygen oxidized (removed some electrons) from paper molecules destroying the bonds that held it together leaving ashes.

In the 1800s physicists, while studying the intensity and wavelength of light emitted by very hot objects, saw patterns in the wavelengths of the emitted light that they could not explain. When we say that a hot object is red hot or white hot, we mean that it is so hot that it is giving off red or white light. In 1900 the German physicist Max Planck explained the patterns in the wavelengths of emitted light by assuming that the light emitted or absorbed by atoms was in set amounts called **quanta** (plural of the singular quantum). He determined that the energy of a quantum was equal to a constant number (called Planck's constant symbolized by the letter "h") times the frequency of the light (symbolized by the Greek letter ν nu). The energy of a quantum of energy is $h\nu$. His ideas fit the data very well and he was awarded the Nobel Prize for his work in 1918. A quantum of energy is so small that it is only observed at the level of an atom. This is why we are not aware of them in our daily lives. The idea that set amounts of energy are absorbed or emitted by atoms is evidence of God's order and design.

In 1905 the physicist Albert Einstein determined that light consisted of packets of energy called **photons** whose energy was $h\nu$. He came to this conclusion by studying the photoelectric effect. This is where light shining on a piece of metal causes the metal to emit electrons. This is the basis of solar panels. He noticed that the light had to be of a certain frequency before the metal would give off any electrons. Lower frequency of light has less energy and higher frequency light has more energy. He reasoned that if light behaved as a wave, the metal could gradually absorb enough energy to emit electrons even if the light was of lower energy (lower frequency). But if the

Albert Einstein and Max Planck in Berlin, 1931.

CHEMISTRY

light were composed of particles (photons) the energy of the photons had to have a minimum amount of energy before any electrons would be emitted and this is what happened. Einstein used Planck's concept and determined that there was a minimum energy (h𝘯) for the metal to give off electrons.

In the 1920s Niels Bohr described electrons as traveling around the nucleus of an atom like satellites around the earth. To allow for the idea of the energy of the electrons being in quanta, he said that the electrons had specific energies or orbits that they could occupy. He called these the orbits of the electrons. His theory was very limited in that it could only predict the behavior of the electrons of a hydrogen atom with only one electron.

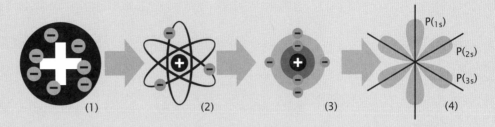

(1) (2) (3) (4)

Ever changing atomic models: Thomson (1), Rutherford (2), Bohr (3), Heisenberg / Schrödinger (4).

This model of the electrons of atoms was later modified by Erwin Schrödinger, who said that instead of orbiting around the nucleus of an atom, electrons had regions where they would more likely be found, depending upon their energy levels that he called orbitals. An orbital is not a place or container. It is a category, such as the cleverest students in class. This opened the door for understanding the behaviors of atoms with multiple electrons. Mainly, it helped predict which bonds would form between which atoms. In other words, it is a major tool to predict the outcomes of chemical reactions.

When you throw a ball, you can throw it at different speeds. A faster-moving ball has more kinetic energy (energy of motion) than a slower-moving ball. Electrons are very different than balls. In fact, they are downright weird. Electrons can only have certain energy levels. Imagine if you could only throw a ball at 10 miles/hour, 20 miles/hour, or at 30 miles/hour with no speeds in between. This means that it cannot go 11 miles/hour. Or imagine a car that goes from 40 miles/hour to 60 miles/hour with no speeds in between. You could get some major whip lash riding in that car. Electrons are like that! To make it even stranger, the energies of electrons are not indicated by their speeds but by their orbitals. An orbital is an indication of an energy level — not physical position. The symbols for the possible orbitals of electrons are

s orbitals

1s

p orbitals

$2p_x$ $2p_y$ $2p_z$

d orbitals

3d   3d   3d   3d   3d

An orbital is an indication of an energy level — not physical position.

written as 1s, 2s, 2p, 3s, 3p, and so on. If you were an electron at the 1s orbital (energy level), you would have less energy than if you were at the 2s orbital. When an electron gains enough energy to go from the 1s to the 2s orbital, it goes directly to the 2s orbital with no energy levels in between. When this was realized, it was probably the most important breakthrough in understanding modern chemistry.

The key to understanding how chemical reactions are predicted is to look at which orbitals of an atom are occupied by electrons. As you will see, this is where the periodic table comes in handy. Orbitals with electrons with lower energies are found in regions closer to the nucleus of an atom. Orbitals with electrons with greater energies are found in regions that are further away from the nucleus of an atom. An important rule to remember is that orbitals of the same energy will each receive 1 electron first and a second electron after they have each received 1. Each orbital can have a maximum of 2 electrons. The available orbitals of an atom are designated as ...

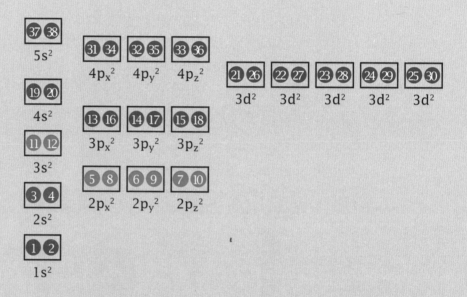

1s is the lowest energy orbital. An orbital indicates the energy of electrons that are in that orbital. Planck and Einstein described the quantum as the amount of energy of an electron, which is the same as in the orbitals.

The s orbitals are always filled one at a time. The p orbitals are filled in groups of 3. The 2p orbitals all have the same energy; the 3p orbitals all have the same energy, and the 4p orbitals all have the same energy. The d orbitals always come in groups of 5. The sequence shown above is important — 1s fills with electrons before 2s, and 2s fills before 2p, which fills before 3s. Notice that the 3d orbitals have more energy than the 4s orbital but less energy than the 4p orbitals. You need to learn this pattern now because you will be using it from here on.

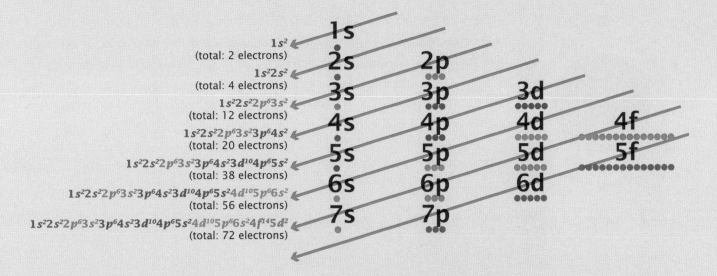

The above Electron Configuration Chart provides a useful way to know how electrons will position themselves in atoms. The red arrows show the path followed as more electrons are added. The last orbital is 7p because there are no known atoms (elements) with more orbitals. This chart shows an example of how to determine the electron configuration of the element with 72 protons and 72 electrons: Hafnium.

The periodic table shows the atomic number of helium (He) to be 2. This means that each helium atom has 2 protons in its nucleus. If the helium atom has neutral charge, it has to have 2 electrons. The **electron configuration** of helium is ...

$$1s^2$$

Helium
$1s^2$

The 2 electrons of the helium atom go into the lowest orbital possible, which is 1s. The 1s orbital can hold 2 electrons so it is written as $1s^2$.

Consider the nitrogen (N) atom. The periodic table shows nitrogen to have an atomic number of 7. With 7 protons, a neutral charged nitrogen atom has 7 electrons to balance the 7 positive charges. The electron configuration of nitrogen is written as ...

$$1s^2 \ 2s^2 \ 2p_x^1 \ 2p_y^1 \ 2p_z^1$$

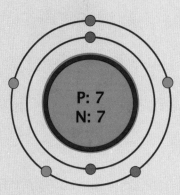

Nitrogen
$1s^2 \ 2s^2 \ 2p_x^1 \ 2p_y^1 \ 2p_z^1$

The 1s and 2s orbitals each have 2 electrons, meaning that they are filled. This leaves 3 out of the 7 electrons to go into the 2p orbitals. Each of the 2p orbitals get 1 electron because they are each at the same energy level.

An oxygen (O) atom has an atomic number of 8 and its electron configuration is ...

$$1s^2 \ 2s^2 \ 2p_x^2 \ 2p_y^1 \ 2p_z^1$$

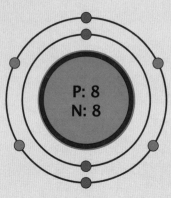

Oxygen
$1s^2\ 2s^2\ 2p_x^2\ 2p_y^1\ 2p_z^1$

After the 1s and 2s orbitals are filled for oxygen, there are 4 out of the 8 electrons left. 1 goes first into each of the 2p orbitals, leaving 1 electron left over. That electron now goes back into the first 2p ($2p_x$) orbital.

You may be wondering where this is going with all this detail. Hang in there and you will see.

Oxygen tends to acquire 2 extra electrons to fill the $2p_y$ and $2p_z$ orbitals.

It now has 2 more negatives (10 electrons) than positive charges (8 protons). Its electron configuration becomes ...

$$1s^2\ 2s^2\ 2p_x^2\ 2p_y^2\ 2p_z^2$$

Oxygen with 2 extra electrons is more stable than an oxygen atom without the 2 extra electrons because the orbitals are more stable when they are filled. Combustion (burning) occurs when oxygen atoms take electrons (which bond atoms together in molecules) away from them, causing them to come apart and form ashes. Oxygen is called a non-metal because one of the properties of non-metals is to attract extra electrons. A property of metals is to readily give up electrons to non-metals. This is why metals are good conductors of electricity and non-metals are not.

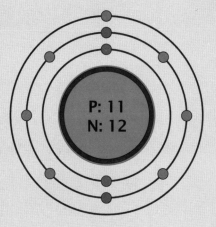

Sodium
$1s^2\ 2s^2\ 2p_x^2\ 2p_y^2\ 2p_z^2\ 3s^1$

Sodium (Na), a metal, has an atomic number of 11 and an electron configuration of ...

$$1s^2\ 2s^2\ 2p_x^2\ 2p_y^2\ 2p_z^2\ 3s^1$$

Sodium with 11 electrons has an electron by itself in the 3s orbital. The negative charged 1s, 2s, and 2p electrons closer to the nucleus act as a shield so that the more energetic 3s electron is not as strongly held by the 11 positive-charged protons in the nucleus. Because of this reduced attraction, sodium atoms tend to readily give up the outer 3s electron to a non-metal atom that may be near. After it loses its 3s electron, the sodium atom becomes Na$^+$ because it now has 11 (+) protons and 10 (-) electrons. Its electron configuration becomes ...

$$1s^2\ 2s^2\ 2p_x^2\ 2p_y^2\ 2p_z^2$$

This is the same as that for O^{--}. Even though they have the same electron configuration, the Na$^+$ and the O^{--} are quite different. Remember the Na$^+$ has 11 protons and the O^{--} has 8 protons. They are still very different elements. These are called sodium and oxygen **ions** because an ion is an atom or molecule whose total charge is positive or negative.

A chlorine atom (Cl) has an atomic number of 17. As a neutrally charged atom, it has 17 electrons with an electron configuration of ...

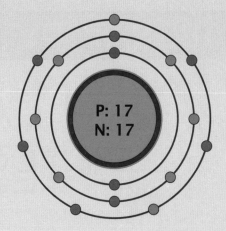

$$1s^2\ 2s^2\ 2p_x^2\ 2p_y^2\ 2p_z^2\ 3s^2\ 3p_x^2\ 3p_y^2\ 3p_z^1$$

Because the 2p orbitals are all filled, they can be written as $2p^6$ giving ...

$$1s^2\ 2s^2\ 2p^6\ 3s^2\ 3p_x^2\ 3p_y^2\ 3p_z^1$$

If you add up the electrons in the Cl atom, you get 17. Notice that the 3p orbitals fill up in the same way as the 2p orbitals with each one getting 1 first and then going back and adding 1 more to each. There is an unpaired $3p_z^1$ electron.

Chlorine
$$1s^2\ 2s^2\ 2p_x^2\ 2p_y^2\ 2p_z^2\ 3s^2\ 3p_x^2\ 3p_y^2\ 3p_z$$
or
$$1s^2\ 2s^2\ 2p^6\ 3s^2\ 3p_x^2\ 3p_y^2\ 3p_z^1$$

To fill all of the orbitals of a Cl atom, it requires 1 more electron. When it takes an electron from a metal atom, it becomes Cl⁻ because now it has 17 (+) protons and 18 (-) electrons. Because Cl attracts an extra electron it is a non-metal. Its electron configuration becomes ...

$$1s^2\ 2s^2\ 2p^6\ 3s^2\ 3p^6$$

Cl⁻ is called a chloride ion instead of chlorine, the name of the element.

When Na and Cl come together, you get Na⁺ and Cl⁻ (table salt). The Cl takes an electron away from the Na and the Na says thank you. To show the significance of the orbitals and the difference that the + and – charges make, consider the sodium and chlorine atoms before they meet. Sodium metal upon contact with water will ignite and burn and possibly have a violent explosion. Chlorine is a green gas that is a deadly poison. Nasty things. Put them together and you have table salt. Other than possibly raising your blood pressure, they are pretty benign as salt. Such a difference a little electron can make.

Positive ions (like Na⁺) form ionic bonds with negative ions (like Cl⁻). These are not actual physical links but sufficiently strong attractions. Most periodic tables have a jagged line separating a few non-metal elements on the right side of the line from the more numerous metal elements on the left side of the line. A helpful rule to remember is that metals tend to lose electrons and non-metals tend to gain electrons.

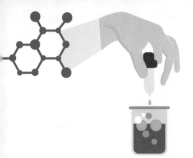

FLAME TESTS

REQUIRED MATERIALS

- Spectroscope Analysis Kit:
  - ~ $CaCl_2$, KCl, $SrCl_2$, LiCl
  - ~ Spectroscope
  - ~ Popsicle sticks

- NaCl (table salt)

- Flame source (Bunsen burner, alcohol burner*, gas stove, oil lamp, or long lighter)

*Important: See "Bunsen/Alcohol Burner Use" in Appendix 1 on page 276

INTRODUCTION

For this exercise, you will need the Spectroscope Analysis Kit. The spectroscope splits light into its various wavelengths. Look through the spectroscope at a source of white light. You should see a rainbow of colors. White light is a mixture of all of the colors. The word *spectrum* refers to something that is lined up from greatest to least. If you had 10 people line up from tallest to shortest, you have a spectrum (also called a gradient). In this case, the colors of light are lined up from greatest energy (violet) to lowest energy (red). The colors from greatest to lowest energy are violet, blue, green, yellow, orange, and red. Red usually looks

brighter to us because our eyes are more sensitive to red than blue (even though blue has more energy).

PURPOSE

This exercise demonstrates the use of emission spectra to identify elements and to visualize the energy levels of their electron orbitals.

PROCEDURE

The following procedures need to be done in a darkened room to keep stray light from interfering with the results.

1. Testing $CaCl_2$

   A. Place a small amount of $CaCl_2$ on the end of one of the wooden sticks from the spectroscope analysis kit.

   B. Place it in a flame so that it glows. It should be a red color.

   C. Observe the flame through the spectroscope.

   Ca^{++} (calcium) has 18 (20 – 2) electrons. Each of the electrons in Ca^{++} have certain energies for its orbitals. When the Ca^{++} is placed in a flame, the electrons get all excited (you would too). They gain energy and go to higher energy orbitals. They are not physically going anywhere. This refers to their gaining set amounts of energy. They then give up that extra energy and return to their original orbitals. The energy that they give off is in the form of light. When you look through the spectroscope, you will see colored lines of light separated by dark spaces. The colored lines of light are the amounts of energy given off by each electron returning to its original energy. The pattern of lines that you see is unique to Ca^{++}. It is called its "fingerprint."

   Write a description of what you observe, using complete sentences. Even though the flame appears to be red, you see more colors with the spectroscope. It is called an emission spectrum because the atoms of Ca++ are emitting a spectrum of light.

   When the wooden stick has cooled down you should place it in water before you dispose of it in the trash.

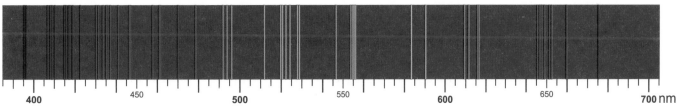

2. Repeat the above procedure using KCl instead.

   In this case, you are seeing the emission spectrum of the K$^+$ (potassium) ion. The flame should be pale violet. Notice that the emission spectrum of K$^+$ is different than that of Ca^{++}. Describe the pattern that you see.

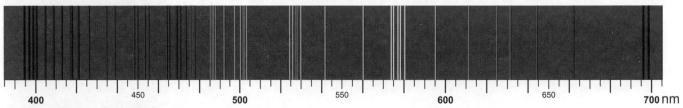

3. Repeat the procedure with SrCl$_2$ (strontium chloride).

   The Sr^{++} ion has 36 (38-2) electrons. The flame from Sr^{++} should appear deep red. Look at and describe its emission spectrum. Again, you should see a pattern that is unique to Sr^{++}.

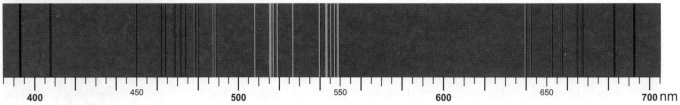

4. Do the procedure with LiCl (lithium chloride).

The Li+ ion has 2 (3-1) electrons. When the Li+ ion is burned it will also give off a deep red flame. Can you tell the flame from Sr++ apart from the flame from Li+? You may not be able to tell them apart, but with the spectroscope their spectra should be different. This is how you can tell them apart.

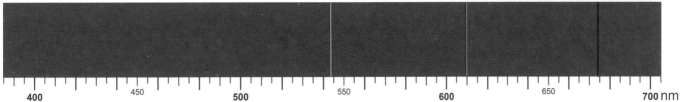

5. Now take some table salt (NaCl) and redo the procedure.

The Na+ ion has 10 (11-1) electrons. You should be getting pretty good at it by now. The flame from the Na+ ion should be yellow. Describe the emission spectrum that you observe for Na+.

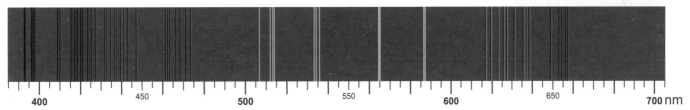

You will be given 2 salts from the above samples. You are to do the flame test on each of them separately and determine their identity. The color of the flame may not be conclusive because the flames for Sr++ and Li+ may look alike. Go by your descriptions of their emission spectra. After you have written out what you think the salts are and why, their identities will be revealed to you. How did you do?

CHAPTER 13

ELECTRON CONFIGURATIONS CONTINUED

OBJECTIVES AND VOCABULARY

The learning objective of lesson thirteen is to give you further practice in writing out the electron configurations of various elements. This is a very important skill that you will use a lot in the following lessons. It is an important tool in understanding the reactions of chemical substances.

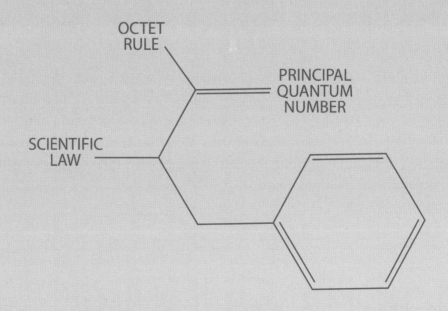

OCTET RULE

PRINCIPAL QUANTUM NUMBER

SCIENTIFIC LAW

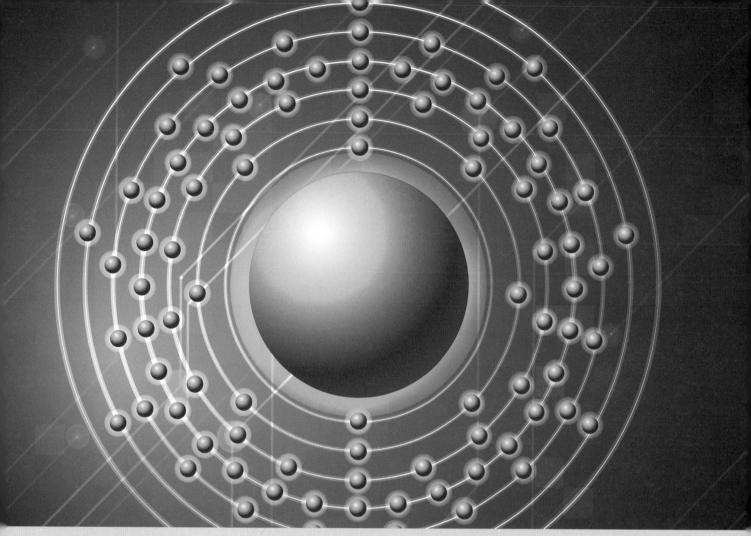

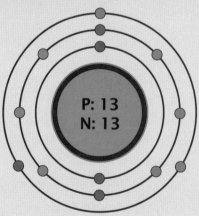

Aluminum
$1s^2\ 2s^2\ 2p^6\ 3s^2\ 3p_x^1$

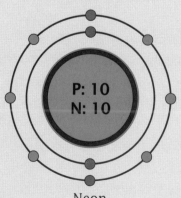

Neon
$1s^2\ 2s^2\ 2p^6$

L ook up Al (aluminum) on the periodic table. Notice that it has an atomic number of 13. You know from practical experience that aluminum is a metal. Besides, it is on the left side of the jagged line on the periodic table. Write out the electron configuration of aluminum.

In order to accommodate the 13 electrons, the electron configuration is ...

$$1s^2\ 2s^2\ 2p^6\ \text{(that is 10 electrons) and}\ 3s^2\ 3p_x^1$$

Now look at the electron configuration of Ne (neon) with 10 electrons.

$$1s^2\ 2s^2\ 2p^6$$

Neon is very stable because it has all of its orbitals filled.

Notice that for aluminum, the highest energy orbitals have the number 3 in front of their electron configurations. For neon, the highest energy electron orbitals begin with 2. This number is

A scientific law *is something that is observed to always occur a certain way.*

called the **principle quantum number.** One of the goals of science is to be able to come up with general principles that are seen in many observations. A **scientific law** is something that is observed to always occur a certain way. It is not a hypothesis or theory because it does not attempt to explain the observations but just to describe them. An example is the Law of Gravity. It is a statement of what happens to objects subjected to forces of gravity. We do not have a good theory of gravity which would be an explanation of gravity. These are hard to account for with a secular model of the universe. Why should matter behave in consistent patterns in and of itself? These are accounted for with a creation model because when God spoke the universe into being, He gave it some laws for which there is no free will. If we try to break these laws it is to our peril. We cannot break them. But they were given to the creation — not to the Creator. So it was no problem for Jesus to walk on water or to raise the dead because He is the Creator — God Himself.

"In the beginning was the Word, and the Word was with God and the Word was God. He was in the beginning with God. All things were made through Him, and without Him was not anything made that was made" John 1: 1,2

As the Creator He as well maintains the creation and enforces the natural laws.

Speaking of Christ … "For by Him all things were created, in heaven and on earth, visible and invisible, whether thrones or dominions or rulers or authorities – all things were created through Him and for Him. And He is before all things, and in Him all things hold together." Colossians 1: 16,17

Octet Rule *states that if the number of electrons with the highest principle quantum number is 8 (an octet), the atom is stable and does not readily react with other atoms.*

A series of observations were made with the electron configurations that led to a rule that holds true (except for ionic compounds of transition metals — do not be concerned about these for now) called the **Octet Rule**. It is called a rule rather than a law because of the exceptions. It states that if the number of electrons with the highest principle quantum number is 8 (an octet), the atom is stable and does not readily react with other atoms. As we saw above, Al has 3 electrons with the highest principle quantum number ($3s^2\ 3p_x^1$) and is reactive as was shown in the examples of balancing equations. However, Ne has 8 electrons with the highest principle quantum number ($2s^2\ 2p^6$) so it is very stable. It is called a noble gas because it almost never reacts.

Consider another example: K (potassium). K has an atomic number of 19. Its electron configuration is $1s^2\ 2s^2\ 2p^6\ 3s^2\ 3p^6\ 4s^1$. The highest principle quantum number is 4 with only 1 electron ($4s^1$). If it lost that electron it would become $1s^2\ 2s^2\ 2p^6\ 3s^2\ 3p^6$. Now its highest principle quantum number is 3 with 8 electrons. That satisfies the Octet Rule. This leaves the K atom with 19 protons and 18 electrons. That is 19 positive charges and 18 negative charges leaving a total of a +1 charge (19 – 18 = 1). It is now written as K^+. Remember that metals tend to lose electrons so that makes K a metal. If you look at the periodic table you can see that it is way to the left of the jagged line.

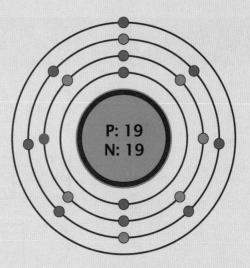

Potassium
$1s^2\ 2s^2\ 2p^6\ 3s^2\ 3p^6\ 4s^1$

Look at Kr (krypton — not kryptonite that Superman fears) as another example. After searching around the periodic table, you stumble across it as element number 36. It is to the right of the jagged line on the periodic table so you would expect it to be a non-metal. Its electron configuration is

$$1s^2\ 2s^2\ 2p^6\ 3s^2\ 3p^6\ 4s^2\ 3d^{10}\ 4p^6$$

If you add these electrons, you get 36. We are looking at this example because it involves the d orbitals. Here the principle quantum number is 4 with 8 electrons ($4s^2\ 4p^6$). Kr is another example along with Ne as a noble gas. It is very stable.

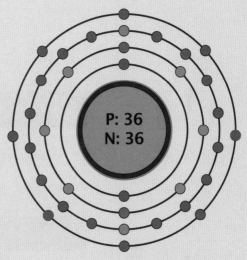

Krypton
$1s^2\ 2s^2\ 2p^6\ 3s^2\ 3p^6\ 4s^2\ 3d^{10}\ 4p^6$

Look at As (arsenic — very poisonous). It has an atomic number of 33. When it has as many positive protons as negative electrons, it has 33 electrons. Its electron configuration is ...

$$1s^2\ 2s^2\ 2p^6\ 3s^2\ 3p^6\ 4s^2\ 3d^{10}\ 4p_x^1\ 4p_y^1\ 4p_z^1$$

Notice that the 4p orbitals are not all filled, so it is broken down as the x, y and z orbitals. In this case the highest principle quantum number is 4. There are 5 electrons in the orbitals with the principle quantum number of 4 ($4s^2\ 4p_x^1\ 4p_y^1\ 4p_z^1$). Therefore, to have 8 electrons with a principle quantum number of 4, it must gain 3 electrons. This gives it the same electron configuration as Kr. Except that it is still As because it has 33 protons and not 36 like Kr. Remember that it is the number of protons (atomic number) that identifies the element. As gains 3 electrons to become As^{-3}. Also remember that in order to gain 3 electrons (become reduced), it has to have some place to take them from (something that becomes oxidized).

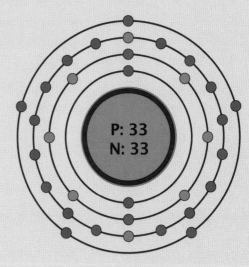

Arsenic
$1s^2\ 2s^2\ 2p^6\ 3s^2\ 3p^6\ 4s^1\ 3d^{10}\ 4p_x^1\ 4p_y^1\ 4p_z^1$

LABORATORY 13

DIAGRAMMING ELECTRON ORBITALS

REQUIRED MATERIALS

- Long balloons
- Masking (or similar) tape

PURPOSE

The shape of electron orbitals indicates where electrons with differing energy levels are more likely to be found around the nucleus of the atom. By diagramming and using balloons, the student will be more familiar with the shape of the orbitals. This will be especially helpful when the chemistry of carbon compounds is studied in later lessons.

PROCEDURE

The s orbitals are drawn as spheres. The 1s orbitals are drawn as a sphere around the nucleus of the atom. For the elements H (1 electron, $1s^1$) and He (2 electrons, $1s^2$), the 1s orbital is …

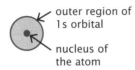

The 2s orbital for the third and fourth electrons of Li (lithium) and Be (beryllium) is drawn as a sphere farther away from the nucleus than the 1s sphere.

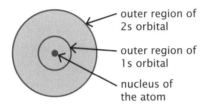

The 2p orbitals are quite different. Blow up a long balloon and twist it in the middle.

This balloon represents one of the 2p orbitals. You need three 2p orbitals ($2p_x$, $2p_y$ and $2p_z$). Blow up 2 more long balloons and twist them in the middle. Now put all 3 balloons together so that the twists are in the middle and the balloons are sticking out in 3 dimensions.

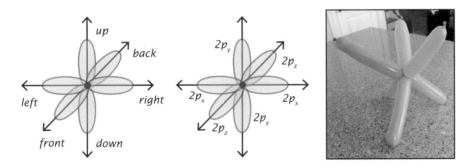

Each balloon has 2 parts. Each 2p orbital has 2 parts (the 2 halves of the same balloon). The 2 electrons that can occupy each 2p orbital are most likely to be found anywhere in the 2 parts of the orbital. The 1s and 2s orbitals would fit in the middle of the twists. Do not try to duplicate that with the balloons.

Take 3 more long balloons and do this again, making another set of 2p orbitals. Use these to represent a second nitrogen atom. A nitrogen atom has an atomic number of 7, meaning that it has 7 electrons if it has not gained or lost any. Its electron configuration is $1s^2\ 2s^2\ 2p_{x1}\ 2p_{y1}\ 2p_{z1}$. Each set of 3 balloons represents the p orbitals of a nitrogen atom. The balloons represent 2 nitrogen atoms each with 3 sets of p orbitals. Each p orbital has 1 electron. Line up the balloons as in this diagram.

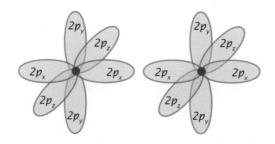

Two balloons directly face each other. Two go up and down vertically and two are horizontal going to the right and left. The two balloons that face each other merge (not literally) and form a covalent bond. Each one has one electron, so together they have 2 electrons shared between each other. The vertical orbitals (balloons) merge and the horizontal orbitals also merge. Using masking (or similar) tape, tape together the $2p_x$ orbitals (facing each other) of both sets of balloons. Tape the $2p_y$ (vertical) orbitals together and the $2p_z$ orbitals (horizontal) together. This gives you three covalent bonds between the nitrogen atoms.

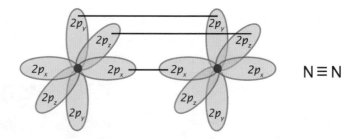

Present your balloon model of the N_2 molecule and a paragraph describing how the 3 shared pairs of orbitals represent the 3 covalent bonds.

Niels Bohr (1885–1962), a Danish physicist, was the first to apply quantum theory to atomic structure. He looked at light from a glowing element (such as hydrogen) through a spectroscope — as you did in a recent lab. The physicist Max Plank (1858–1947) proposed that energy (such as light) traveled through space in bundles called quanta (singular is quantum). This was applied by Bohr as quantum theory.

Discoveries *in* CHEMISTRY

He identified the lines of a light spectrum with what he called electron orbits.

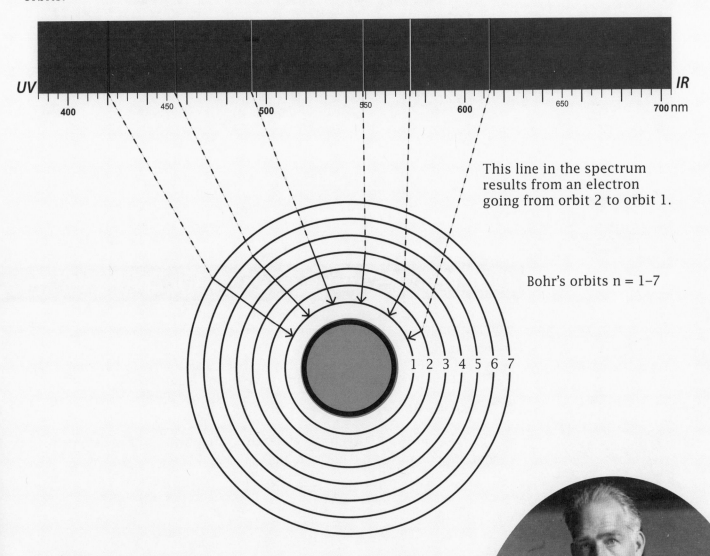

UV   IR

400   450   500   550   600   650   700 nm

This line in the spectrum results from an electron going from orbit 2 to orbit 1.

Bohr's orbits n = 1–7

1 2 3 4 5 6 7

He said that when an electron goes from orbit 2 to orbit 1, it gives off a quantum of light energy as shown by the far right-hand line of the emission spectrum. When an electron went from orbit 7 to orbit 1, it gave off a quantum of light energy as shown by the far left-hand line of the emission spectrum.

Niels Bohr (1935)

CHAPTER 14

PERIODIC TABLE OF THE ELEMENTS

OBJECTIVES AND VOCABULARY

The learning objectives of this lesson are to be able to demonstrate an understanding of:

1. The development of the periodic table of the elements

2. The similarities of the elements in each group

3. The relationship between the electron configurations of the elements and the periods of the periodic table.

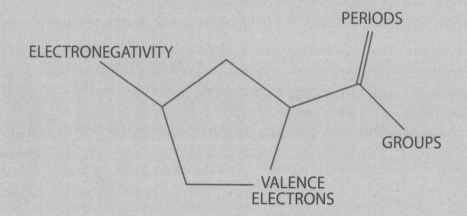

Monument to the periodic table, in front of the Faculty of Chemical and Food Technology of the Slovak University of Technology in Bratislava, Slovakia. The monument honors Dmitri Mendeleev.

The periodic table is a chart with the elements that have similar properties grouped together. Their properties depend a great deal upon their electrons so the table groups them together by similar electron configurations. This is why so much attention was given to them in earlier lessons.

In 1864, John Newlands reported that the elements could be arranged in patterns so that those that would undergo similar chemical reactions would be grouped together. Dmitri Mendeleev arranged the known elements in order of increasing atomic mass. He placed those with similar chemical and physical properties together in the same columns. Later in 1869, he found that there were gaps in the pattern. The gaps between the elements were similar to this example. Suppose you have these numbers arranged in a row 5, 10, 15, 25, 30, and 40. The pattern is that the numbers are increasing by 5. The numbers 20 and 35 are missing. When Mendeleev listed the elements by their atomic masses, he found that some were missing in the pattern. The gaps represented elements that were yet to be discovered. As years went by, these elements were discovered because the patterns seen by Mendeleev directed their search.

PERIODIC TABLE OF THE ELEMENTS

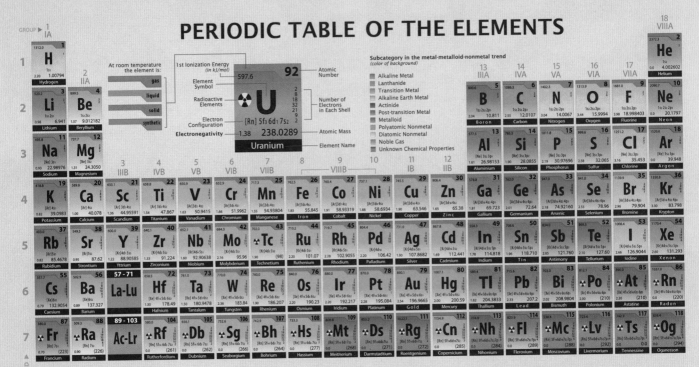

Note: There is a removable periodic table poster in the back of this book.

The vertical columns of the periodic table are called **Groups** and the horizontal rows are called **Periods**. The elements that are in the same Group can be expected to undergo similar chemical reactions. The most important parts of the periodic table are the Groups. They are numbered from 1 through 18. You can also see the older designations of the groups as IA, IB, etc. The Groups 3 through 12 are metals with partially filled d orbitals, also called the Transitional Metals.

The electron configurations of the first 4 elements of Group 1 are ...

Period	Element	Atomic Number	Electron Configuration
1	H	1	$1s^1$
2	Li	3	$1s^2\ 2s^1$
3	Na	11	$1s^2\ 2s^2\ 2p^6\ 3s^1$
4	K	19	$1s^2\ 2s^2\ 2p^6\ 3s^2\ 3p^6\ 4s^1$

Look at the electron configurations of these elements in Group 15.

Period	Element	Atomic Number	Electron Configuration
2	N	7	$1s^2\ 2s^2\ 2p_x^1\ 2p_y^1\ 2p_z^1$
3	P	15	$1s^2\ 2s^2\ 2p^6\ 3s^2\ 3p_x^1\ 3p_y^1\ 3p_z^1$
4	As	33	$1s^2\ 2s^2\ 2p^6\ 3s^2\ 3p^6\ 4s^2\ 3d^{10}\ 4p_x^1\ 4p_y^1\ 4p_z^1$

After carefully looking at these examples, what do the elements in each group have in common with each other? Hint–look at their s orbitals (Group1) and p orbitals (Group 15).

Look at the examples in Group 1. Will they tend to gain or lose electrons? They will lose electrons because they each have a half-filled orbital and they are metals.

How many electrons will they lose? One.

How do you know? They only need to lose 1 electron to leave filled orbitals. You can also use the Octet Rule to come up with 8 electrons in the highest principle quantum number except for Periods 1 and 2 because they have fewer electrons. The electrons with the highest principle quantum number are called **valence electrons**.

Look at the examples from Group 15. Will they tend to gain or lose electrons? They will gain electrons to fill their p orbitals.

How many electrons will they gain? They will each gain 3 electrons to fill their p orbitals and have 8 valence electrons satisfying the Octet Rule.

Notice that the outer electrons for the elements in Group 1 are ...

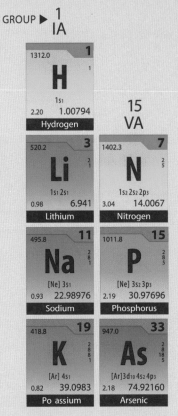

$1s^1$ for H;
$2s^1$ for Li;
$3s^1$ for Na and
$4s^1$ for K

Notice that the outer electrons for the elements in Group 2 are ...

$1s^2$ for He;
$2s^2$ for Be;
$3s^2$ for Mg and
$4s^2$ for Ca

Notice that the outer electrons for the elements in Group 13 are ...

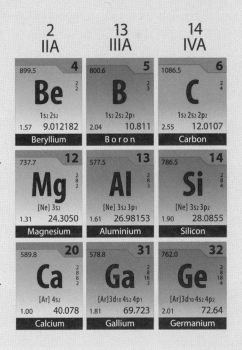

$2p_x^1$ for B;
$3p_x^1$ for Al and
$4p_x^1$ for Ga

Notice that the outer electrons for the elements in Group 14 are ...

$2p_x^1 2p_y^1$ for C;
$3p_x^1 3p_y^1$ for Si and
$4p_x^1 4p_y^1$ for Ge

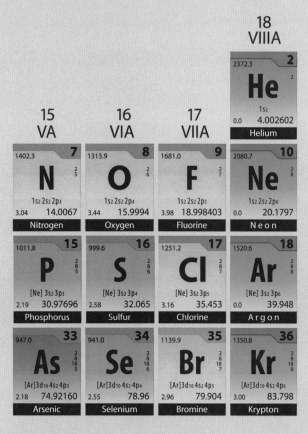

18
VIIIA

2372.3	**2**
He	₂
1s2	
0.0	4.002602
Helium	

15
VA

16
VIA

17
VIIA

1402.3	**7**	1313.9	**8**	1681.0	**9**	2080.7	**10**
N	2 5	**O**	2 6	**F**	2 7	**Ne**	2 8
1s2 2s2 2p3		1s2 2s2 2p4		1s2 2s2 2p5		1s2 2s2 2p6	
3.04	14.0067	3.44	15.9994	3.98	18.998403	0.0	20.1797
Nitrogen		Oxygen		Fluorine		Neon	

1011.8	**15**	999.6	**16**	1251.2	**17**	1520.6	**18**
P	2 8 5	**S**	2 8 6	**Cl**	2 8 7	**Ar**	2 8 8
[Ne] 3s2 3p3		[Ne] 3s2 3p4		[Ne] 3s2 3p5		[Ne] 3s2 3p6	
2.19	30.97696	2.58	32.065	3.16	35.453	0.0	39.948
Phosphorus		Sulfur		Chlorine		Argon	

947.0	**33**	941.0	**34**	1139.9	**35**	1350.8	**36**
As	2 8 18 5	**Se**	2 8 18 6	**Br**	2 8 18 7	**Kr**	2 8 18 8
[Ar]3d10 4s2 4p3		[Ar]3d10 4s2 4p4		[Ar]3d10 4s2 4p5		[Ar]3d10 4s2 4p6	
2.18	74.92160	2.55	78.96	2.96	79.904	3.00	83.798
Arsenic		Selenium		Bromine		Krypton	

Notice that the outer electrons for the elements in Group 15 are ...

$2p_x^1 2p_y^1 2p_z^1$ for N;
$3p_x^1 3p_y^1 3p_z^1$ for P and
$4p_x^1 4p_y^1 4p_z^1$ for As

Notice that the outer electrons for the elements in Group 16 are ...

$2p_x^2 2p_y^1 2p_z^1$ for O;
$3p_x^2 3p_y^1 3p_z^1$ for S and
$4p_x^2 4p_y^1 4p_z^1$ for Se

Notice that the outer electrons for the elements in Group 17 are ...

$2p_x^2 2p_y^2 2p_z^1$ for F;
$3p_x^2 3p_y^2 3p_z^1$ for Cl and
$4p_x^2 4p_y^2 4p_z^1$ for Br

Notice that the outer electrons for the elements in Group 18 the Noble Gas Group are ...

$2p^6$ for Ne;
$3p^6$ for Ar and
$4p^6$ for Kr

The filled orbitals for the Noble Gases are a very stable arrangement. These elements are very unreactive. They used to be called Inert Gases because they had not been observed to have reacted with anything. But they are now called Noble Gases because they have been found to engage in very limited chemical reactions. With 6 electrons in their outer p orbitals, it is very difficult to remove any electrons and they have no unpaired electrons that have to move to any other orbitals. Remember that the Group 2 elements' outer electrons fill an s orbital. For example, the electron configuration of Ca ends with $4s^2$. But even though the 4s orbital is filled, it still loses the $4s^2$ electrons to become Ca⁺⁺. The $4s^2$ electrons of Ca are not held by the nucleus as strongly as the $2p^6$ electrons of Ne because the $4s^2$ electrons of Ca are shielded from the nucleus by the $1s^2$ $2s^2$ $2p^6$ $3s^2$ and $3p^6$ electrons. This enables Ca to have a more stable state when it has the same electron configuration as the noble gas Ar. Ca⁺⁺ and Ar both have the electron configuration $1s^2$ $2s^2$ $2p^6$ $3s^2$ $3p^6$.

The filled orbitals for the Noble Gases are a very stable arrangement.

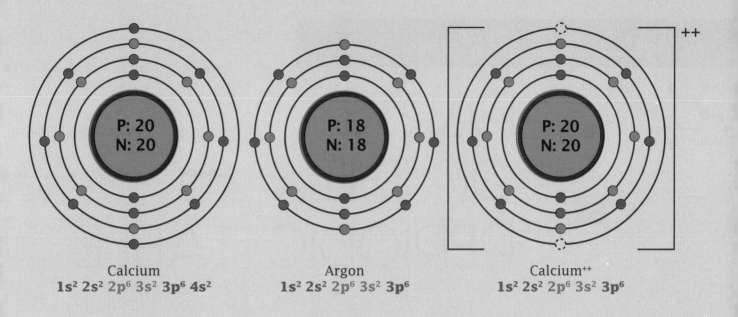

Calcium
$1s^2\ 2s^2\ 2p^6\ 3s^2\ 3p^6\ 4s^2$

Argon
$1s^2\ 2s^2\ 2p^6\ 3s^2\ 3p^6$

Calcium++
$1s^2\ 2s^2\ 2p^6\ 3s^2\ 3p^6$

SCHRÖDINGER'S WAVE MECHANICS THEORY

Niels Bohr was only able to predict the energy values of the one electron of a hydrogen atom. His theory did not work for atoms with more than one electron. The quanta that Bohr used referred to the light emitted from glowing elements.

Discoveries *in* **CHEMISTRY**

Erwin Schrödinger

Louis de Broglie stated in 1924 that electrons should move in wave patterns. Erwin Schrödinger applied this concept in predicting the energies of electrons with what he called a wave function (a mathematical concept). From the wave function, he came up with the probabilities of finding electrons in regions he called orbitals. The shapes of the balloons you used in a recent lab exercise demonstrate the orbitals where you are most likely to find the electrons. These replaced Niels Bohr's orbits.

Even though Bohr's theory was limited, it gave the background that Schrödinger could build upon.

PARTS OF THE PERIODIC TABLE

REQUIRED MATERIALS

- Periodic table
- Blank periodic table

PURPOSE

This exercise is designed to reinforce the student's knowledge of the periodic table.

PROCEDURE

It was stated in earlier lessons that it was not necessary to memorize the periodic table. It is a place where you look things up. In order to do so, you need to know your way around the periodic table. Take a blank periodic table (with empty squares) and fill it in as directed. Be careful to be neat and legible.

At the end of this exercise you will have a periodic table with the atomic numbers, element names, and electron configuration for the first 36 elements filled in.

1. In the right-hand corner in the top of the boxes (up to element 36), write in the atomic numbers using another periodic table for the information.

2. For number 1 (H) write in $1s^1$.

3. For element number 2 (He) write in $1s^2$.

These are the filled orbitals for these elements.

4. For element 3 (Li) write in $1s^2\ 2s^1$.

5. For element 4 (Be) write in $1s^2\ 2s^2$.

6. For element 5 (B) write in $1s^2\ 2s^2\ 2p_x^1$.

7. For element 6 (C) write in $1s^2\ 2s^2\ 2p_x^1\ 2p_y^1$.

8. For element 7 (N) write in $1s2\ 2s^2\ 2p_x^1\ 2p_y^1\ 2p_z^1$.

9. For element 8 (O) write in $1s^2\ 2s^2\ 2p_x^2\ 2p_y^1\ 2p_z^1$.

10. For element 9 (F) write in $1s^2\ 2s^2\ 2p_x^2\ 2p_y^2\ 2p_z^1$.

11. For element 10 (Ne) write in $1s^2\ 2s^2\ 2p^6$.

12. For element 11 (Na) write in $1s^2\ 2s^2\ 2p^6\ 3s^1$.

13. For element 19 (K) write in $1s^2\ 2s^2\ 2p^6\ 3s^2\ 3p^6\ 4s^1$.

Look over the pattern that you have produced in the periodic table.

14. Using that pattern, fill in the electron configurations for elements 12 through 18.

15. Fill in the electron configurations for elements 20–36.

The electron configurations of elements 21 (Sc) through 30 (Zn) end in $4s^2\ 3d^1$; $4s^2\ 3d^2$; $4s^2\ 3d^3$ through $4s^2\ 3d^{10}$. For elements 31 (Ga) through 36 (Kr), the electron configurations end in $4p_x^1$ through $4p^6$. You have

probably noticed that you need to remember what you studied in this week's lesson to do this exercise.

Look over your filled-in periodic table. Notice that the 2 columns on the left-hand side are the s orbitals. The 6 columns on the right-hand side are the p orbitals. Remember that the columns of the periodic table are the groups. The elements in the same group are similar to each other because their electron configurations end the same way. The rows are the periods. Elements 1, 3, 11, 19, and 37 are in Group 1. Their electron configurations end in $1s^1$, $2s^1$, $3s^1$, $4s^1$, and $5s^1$. Whether they gain or lose or share electrons is determined by their electron configuration. Elements 7, 15, 33, and 51 are in Group 15. Their electron configurations end in $2p_x^1\,2p_y^1\,2p_z^1$; $3p_x^1\,3p_y^1\,3p_z^1$; $4p_x^1\,4p_y^1\,4p_z^1$ and $5p_x^1\,5p_y^1\,5p_z^1$.

PERIODIC TABLE OF THE ELEMENTS

1	2	3	4	5	6	7	8	9	10	11	12	13	14	15	16	17	18
H 1 $1s^1$																	**He** 2 $1s^2$
Li 3 $1s^2 2s^1$	**Be** 4 $1s^2 2s^2$											**B** 5 $1s^2 2s^2 2p_x^1$	**C** 6 $1s^2 2s^2\,2p_x^1 2p_y^1$	**N** 7 $1s^2 2s^2\,2p_x^1 2p_y^1 2p_z^1$	**O** 8 $1s^2 2s^2\,2p_x^2 2p_y^1 2p_z^1$	**F** 9 $1s^2 2s^2\,2p_x^2 2p_y^2 2p_z^1$	**Ne** 10 $1s^2 2s^2 2p^6$
Na 11 $1s^2 2s^2 2p^6\,3s^1$	**Mg** 12 $1s^2 2s^2 2p^6\,3s^2$											**Al** 13 $1s^2 2s^2 2p^6\,3s^2 3p_x^1$	**Si** 14 $1s^2 2s^2 2p^6 3s^2\,3p_x^1 3p_y^1$	**P** 15 $1s^2 2s^2 2p^6 3s^2\,3p_x^1 3p_y^1 3p_z^1$	**S** 16 $1s^2 2s^2 2p^6 3s^2\,3p_x^2 3p_y^1 3p_z^1$	**Cl** 17 $1s^2 2s^2 2p^6 3s^2\,3p_x^2 3p_y^2 3p_z^1$	**Ar** 18 $1s^2 2s^2 2p^6\,3s^2 3p^6$
K 19 $1s^2 2s^2 2p^6\,3s^2 3p^6 4s^1$	**Ca** 20 $1s^2 2s^2 2p^6\,3s^2 3p^6 4s^2$	**Sc** 21 $1s^2 2s^2 2p^6\,3s^2 3p^6 4s^2 3d^1$	**Ti** 22 $1s^2 2s^2 2p^6\,3s^2 3p^6 4s^2 3d^2$	**V** 23 $1s^2 2s^2 2p^6\,3s^2 3p^6 4s^2 3d^3$	**Cr** 24 $1s^2 2s^2 2p^6\,3s^2 3p^6 4s^2 3d^4$	**Mn** 25 $1s^2 2s^2 2p^6\,3s^2 3p^6 4s^2 3d^5$	**Fe** 26 $1s^2 2s^2 2p^6\,3s^2 3p^6 4s^2 3d^6$	**Co** 27 $1s^2 2s^2 2p^6\,3s^2 3p^6 4s^2 3d^7$	**Ni** 28 $1s^2 2s^2 2p^6\,3s^2 3p^6 4s^2 3d^8$	**Cu** 29 $1s^2 2s^2 2p^6\,3s^2 3p^6 4s^2 3d^9$	**Zn** 30 $1s^2 2s^2 2p^6\,3s^2 3p^6 4s^2 3d^{10}$	**Ga** 31 $1s^2 2s^2 2p^6\,3s^2 3p^6 4s^2 3d^{10}\,4p_x^1$	**Ge** 32 $1s^2 2s^2 2p^6\,3s^2 3p^6 4s^2 3d^{10}\,4p_x^1 4p_y^1$	**As** 33 $1s^2 2s^2 2p^6 3s^2\,3p^6 4s^2 3d^{10}\,4p_x^1 4p_y^1 4p_z^1$	**Se** 34 $1s^2 2s^2 2p^6\,3s^2 3p^6 4s^2 3d^{10}\,4p_x^2 4p_y^1 4p_z^1$	**Br** 35 $1s^2 2s^2 2p^6\,3s^2 3p^6 4s^2 3d^{10}\,4p_x^2 4p_y^2 4p_z^1$	**Kr** 36 $1s^2 2s^2 2p^6\,3s^2 3p^6 4s^2 3d^{10}\,4p^6$

Your report for this exercise is the blank periodic table from the Teacher Guide with the electron configurations for elements 1 through 36 filled in. A few have been filled in for you to show how to do them.

Write a sentence or two explaining why the elements in the same group have similar properties.

In 1864, the English chemist John Newlands observed that when the elements were placed in order of increasing atomic weight, every eighth element had similar properties. At this time, they had no knowledge of atomic structures, atomic numbers, and electron configurations. This was a pattern that he observed that he called the Law of Octaves. Most felt that his work had no merit and someone mocked him, saying that he should arrange them alphabetically. In 1869, Dmitri Mendeleev presented a paper showing a periodic table based upon the atomic weights. He said that there were some gaps because of larger differences between some of the atomic weights. In 1882, Mendeleev was awarded the highest award from the Royal Society in London called the Davy Award. Five years later, Newlands was also awarded a Davy Award for his contributions. This shows that major discoveries are the work of many individuals and that sometimes the value of someone's work is not recognized right away.

Not all of the major discoveries in chemistry were done by Christians. But God uses many different individuals to still accomplish His purpose. He has gradually revealed His creation to us as time has gone by.

Discoveries *in* **CHEMISTRY**

Dmitri Mendeleev in the 1890s

A handwritten version of the elements system (Mendeleev's periodic law), based on atomic weight and chemical resemblance (1869).

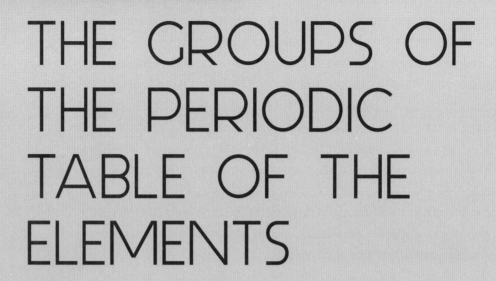

CHAPTER 15

THE GROUPS OF THE PERIODIC TABLE OF THE ELEMENTS

OBJECTIVES AND VOCABULARY

The learning objectives of this lesson are to gain a basic understanding and knowledge of:

1. Some of the more commonly used Groups of the periodic table

2. Some of the more commonly encountered elements.

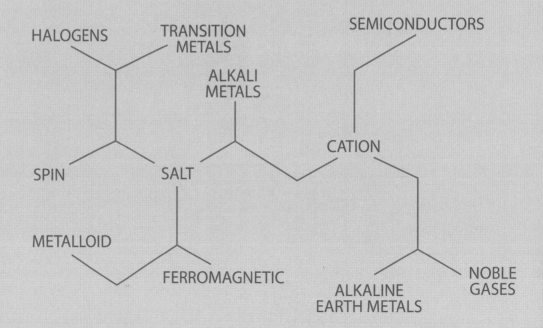

Sodium metal explosion on water.

An extensive study of the groups and elements of the periodic table could take more than a lifetime. This lesson covers some of the more frequently encountered elements.

The first distinction is between metals and non-metals. The metals are found to the left of the jagged line and comprise over 75 percent of the elements. Metal atoms tend to lose some of their valence electrons and are good conductors of electricity and heat. Non-metal atoms tend to gain electrons adding to their valence electrons and are poor conductors of electricity and heat. This is why rubber makes a good insulation around electric wires, and wooden handles on pans heated on the stove are safer than metal handles.

Group 1 (IA) metals are called the **Alkali Metals**. The word "alkali" comes from the Arabic word for "ashes." The term is used because early chemists isolated compounds containing sodium and potassium from wood ashes. Sodium and potassium are very important for the conduction of nerve impulses and muscle contractions. Na$^+$ and K$^+$ are important electrolytes in body

Pieces of potassium metal

NGC 604, a giant region of ionized hydrogen in the Triangulum Galaxy.

fluids. Plants require potassium for growth and development. The Group 1 metals are soft enough to cut with a knife. They are usually softer on the outside and harder on the inside.

The Group 1 elements all end in s^1 in their electron configurations so they tend to lose the 1 electron ending up with a +1 charge.

The first element in this group is hydrogen with 1 proton and 1 electron. This is the most abundant element in the universe. The second most abundant element is He (helium). The stars, including our Sun, are 90 percent H and 10 percent He and a very small percent of the other elements. The electron configuration of H is $1s^1$. It is placed in Group 1 because it has the s^1 electron. Having only 1 electron it does not behave like the other elements in this group. Under great pressures, like in the interior of the large gaseous planets Jupiter, Saturn, Uranus, and Neptune, it behaves like a metal. H can be bonded to a non-metal, such as chloride and sulfur. With chloride it has a +1 charge forming hydrochloric acid (HCl).

$$HCl \rightarrow H^+ \text{ (acid)} + Cl^- \text{ (chloride)}$$

With sulfur, it forms H_2S (hydrogen sulfide), which is poisonous and has rotten egg odor.

Two hydrogen atoms can bond together to form explosive H_2 gas.

When an electric current is passed through water, water comes apart to yield H_2 and O_2 gases.

$$2 \, H_2O \rightarrow 2 \, H_2 + 2 \, O_2$$

They can be very reactive (release enough heat to explode or ignite in flames) with water. A few years ago at a local landfill, the garbage ignited into flames when it rained. Isn't water supposed to put out fires rather than start them? In this case, someone had thrown some sodium metal into the trash and it was taken to the landfill.

When discovered in 1817, Li (lithium) was thought of as having healing powers. This is a good example of why it is important to conduct experiments and test ideas. Sometimes what seems obvious may not be true. Li was used as lithium citrate in the soft drink Seven-Up when it first came out. It was later taken out in the early 1950s. It has since been demonstrated that lithium is effective in smoothing out mood swings with bipolar affective disorder and manic depressive illnesses.

Group 2 (2A) elements are the **Alkaline Earth Metals**. They all end in s^2 in their electron configurations, so they tend to lose the 2 s^2 valence electrons, ending up with a +2 charge, leaving them with 2 more protons than electrons. They are harder and less reactive than the Group 1 elements. They react with water to form hydroxides. An example is Ca (calcium).

$$Ca\ (s) + 2\ H_2O\ (l) \rightarrow Ca(OH)_2\ (aq) + H_2\ (g)$$

Calcium grains, with each grain about 1 mm.

The symbol (s) means solid, (l) means liquid, (aq) means aqueous or dissolved in water, and (g) means gas. The OH^- is called a hydroxide.

Ca (calcium) and Mg (magnesium) are essential for living organisms.

Group 16 (6A) is the Oxygen Group. These elements have 4 electrons in their p orbitals ($p_x^2\ p_y^1\ p_z^1$) and gain 2 more electrons to give them 8 valence electrons, satisfying the Octet Rule. Oxygen, sulfur, and selenium are strong oxidizers (electron thieves) as non-metals.

Te (tellurium) is just above the jagged line. It is called a **metalloid** because its properties are between those of a metal and a non-metal. Si (silicon) and Ge (germanium) as metalloids are called **semiconductors** because they weakly conduct electric currents. They form integrated circuits acting like atomic sized "wires" to conduct miniature electric currents. They are also used widely in computer chips. Their electrical conductance is much less than metals. Te is a metalloid but not a good semiconductor.

Computer chip electronic circuit board.

Ozone (O_3) is a toxic compound with a pungent odor. It is found in small amounts in the upper atmosphere, acting as a shield to protect us from ultra-violet radiation from the Sun. It is also found in air pollution. It can be formed by the high temperature of lightning.

$$3\ O_2\ (g) \rightarrow 2\ O_3\ (g)$$

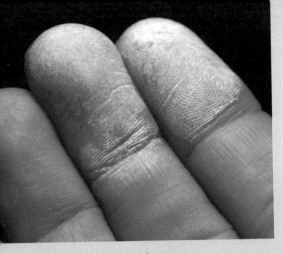

Fingertips shortly after a brief exposure to 35 percent hydrogen peroxide

Hydrogen peroxide (H_2O_2) is widely used as a disinfectant. It breaks down to form water and oxygen. That is the foaming that it produces.

$$2\ H_2O_2\ (aq) \rightarrow 2\ H_2O\ (l) + O_2\ (g)$$

The cap of a bottle of hydrogen peroxide allows for O_2 gas to escape. H_2O_2 is a strong oxidizer because of the O_2 that it releases. H_2O_2 purchased over the counter is 3 percent H_2O_2. It can be dangerous at 30 percent, which is used at times in chemistry labs.

S (sulfur) forms S^{--}, which combines with metal ions to form black sulfide compounds.

$$Mg^{++} + S^{--} \rightarrow MgS\ (magnesium\ sulfide)$$

As described earlier, sulfur combines with hydrogen to form hydrogen sulfide.

$$2\ H^+ + S^{--} \rightarrow H_2S$$

S combines with O_2 to form SO_2 (sulfur dioxide), which is a component in air pollution.

Group 17 (7A) Halogens have 5 electrons in their p orbitals ($p_x^2\ p_y^2\ p_z^1$). They will readily acquire 1 electron to complete their p orbitals satisfying the Octet Rule. The word halogen comes from the Greek words *halos* and *gennao* meaning salt forms. A **salt** is a **cation** (positive charged ion) and a halogen with a –1 charge. The most common example is NaCl sodium chloride – table salt. But KCl (potassium chloride) and $MgCl_2$ (magnesium chloride) are also salts. Sometimes people use a salt substitute (KCl) when they are on a low sodium diet. Another example of a salt is sodium bicarbonate ($NaHCO_3 \rightarrow Na^+ + HCO_3^-$). A salt can also be defined as the result when a metal ion is substituted for the H^+ in an acid. An example could be where Na^+ is substituted for the H^+ in acetic acid (vinegar). In this example an anion (negative charged ion) other than a halogen is used to form a salt.

$$NaCl + CH_3COOH \rightarrow CH_3COONa + HCl$$

CH_3COO^- is the acetate ion

Group 18 (8A) is called the **Noble Gases**. They have all of their p orbitals filled satisfying the Octet Rule without having to gain or lose any electrons. Until the early 1960s they were called the Inert Gases. Some molecules were formed with them using strong oxidizers that could remove 1 of the p electrons. These molecules are only stable at very low temperatures. These Noble Gases include He (helium), Ne (neon), Ar (argon), Kr (krypton), Xe (xenon) and Rn (radon).

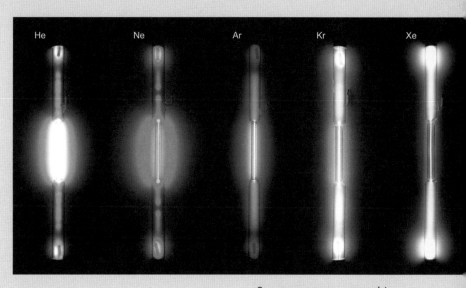

Spectrum represented in gas discharge tubes. The noble gases: helium (He), neon (Ne), argon (Ar), krypton (Kr), xenon (Xe).

Helium is considered to be the most important Noble Gas. It is the second most abundant element in the universe with hydrogen being the most abundant. It has the lowest boiling point of all of the elements. It boils at about 4 degrees above absolute zero. It also has the most unusual history. It is the only element to be discovered on the sun. How was helium discovered on the sun when we cannot take samples of the sun? We went to the sun at night. Right? No, it was done by using the emission spectrum from the light emitted from the glowing gases of the sun. When light from a hot glowing gas (like hydrogen or helium) is passed through a diffraction grating, the light is split into its separate wavelengths. This is what happens when light is passed through droplets of water in the sky producing a rainbow. Imagine a rainbow with most of it missing and only a few colored lines visible. A pattern was seen when sunlight was passed through a diffraction grating (piece of opaque material with many very narrow slits through which light can pass) and some of the wavelengths were not seen before with any of the hot glowing gases used here on earth. They were from an element unknown on earth up to this point. They named it helium after the Greek word *helios* for sun. Helium was later discovered to be abundant in natural gas wells.

The **Transition Metals** are **Groups 3–12 (3B–12B)**. They represent the filling of the d orbitals. They do not all follow the Octet Rule. The pattern of filling their orbitals is quite different than the lighter elements. Elements 57–70 fit in between elements 56 and 71. Element 56 is Ba (barium) and has 2 electrons in its 6s orbital. Before the 5d orbitals begin to fill, the 4f orbitals are filled with elements 57–70. These are called the Lanthanide Series. Then the 5d orbitals are filled with elements 71–80. In the next period, the elements 89–102 have their 5 f orbitals filled (called

> *God's creation is very orderly and He maintains it as such. But at the same time it reflects His authority over creation and not always how we would like it to be. It reflects His great wisdom, which is greater than ours.*

the Actinide Series) and afterward the elements 103–112 have their 6d orbitals filled.

In looking at these odd patterns, one may think that it does not make sense. Why would you have 6s orbitals followed by 4f orbitals, then 5d orbitals and 6p orbitals? These are not based upon the kind of nice neat packages that we would like. They are based upon the physical and chemical properties of the elements as God created them. God's creation is very orderly, and He maintains it as such. But at the same time, it reflects His authority over creation and not always how we would like it to be. It reflects His great wisdom, which is greater than ours. The more we learn and do research, the more we realize what we have yet to understand and learn. His understanding is far above ours. We can see His character and wisdom in the study of His creation. It should cause us to worship Him all the more.

Gold nugget

Au (gold) number 79 is an example of a transition metal found in Group 11 (11B). Gold is valuable because of its rarity and beauty. It is fairly soft compared to the other metals. This makes it easier to make it into jewelry. It is quite resistant to acids. Because of its stability it is used in dentistry. It is combined with other metals to make alloys to make it harder.

Fe (iron) is element number 26 in Group 8 (8B) of the transition metals. It has 6 electrons in its 3d orbitals. Iron is **ferromagnetic,** meaning that iron atoms are tiny magnets. This is because iron has an unpaired electron. Electrons have a property called **spin**. When a charged object, like an electron spins, it produces a magnetic field. In atoms with its electrons paired (remember that each orbital holds 2 electrons), the electrons spin in opposite directions producing magnetic fields in opposite directions that cancel out each other. If an atom has an unpaired electron because an orbital is half filled, its magnetic field is not cancelled out. When an iron object is placed near or touches a strong magnet, its magnetic fields are lined up to match that of the external magnet. Now that the magnetic fields of the iron atoms are all pointing in the same direction, they add to each other producing a strong magnet. If you take the blade of a screw driver and bring it near or touch a strong magnet with it, the blade of the screw driver will become a permanent magnet.

Iron is an important ingredient in the middle of a hemoglobin complex in blood. The ferromagnetic iron atom attracts an O_2 molecule so that the hemoglobin can transport the O_2 from the lungs to the cells of a body.

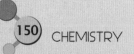

Iron is commonly found dissolved in water as the Fe^{++} (ferrous ion) and the Fe^{+++} (ferric ion).

Together with Mg and Ca, iron contributes to the hardness of water. When you have hard water (usually from wells), it leaves deposits and makes it hard to get soap and shampoo to lather. Sometimes homes with galvanized pipes and hard water build up deposits and have to be repiped.

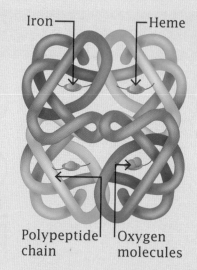

Diagram of hemoglobin

Iron has an interesting property in that it is an upper limit for producing new elements by nuclear fusion (combining their nuclei). An iron atom has 26 protons. When attempts are made to combine nuclei of atoms producing more than a combined number of 26 protons, nuclear fission occurs blowing the atom apart. If the universe began as depicted in the Big Bang Model, the periodic table should end at iron. But that would not even be half of the elements that we have. The only conceivable means whereby we could have all the elements that we have is for God to have created them in the first place. That would not fit with the Big Bang. It is easier to believe what God said in the first place. An important principle comes out of this in that when good honest impartial research is done, the results more and more come to line up with what God said to begin with.

If the universe began as depicted in the Big Bang Model, the periodic table should end at iron. But that would not even be half of the elements that we have.

GROUPS OF THE PERIODIC TABLE

REQUIRED MATERIALS

- Spectroscope Analysis Kit

- Table salt

- Epsom salts

- TUMS®

- Alcohol burner

- KCl (salt substitute)

PURPOSE

This exercise is to help the student remember various properties of some of the elements and groups of the periodic table.

PROCEDURE

The following flame test procedures need to be done in a darkened room to keep stray light from interfering with the results.

The Group 1 elements are the Alkali Metals. These include Li (lithium), Na (sodium), K (potassium), Rb (rubidium), Cs (cesium), and Fr (francium). Of these, the common ones are Na and K. Take a sample of NaCl and KCl (called a salt substitute). Taste each one and indicate whether or not you can tell the difference. In Lab Exercise #12, you tested elements by a flame test. Do this for NaCl and KCl. Can you see a difference in the flame test between them? Look through the spectroscope at the flame of each. Can you tell them apart? How are they similar?

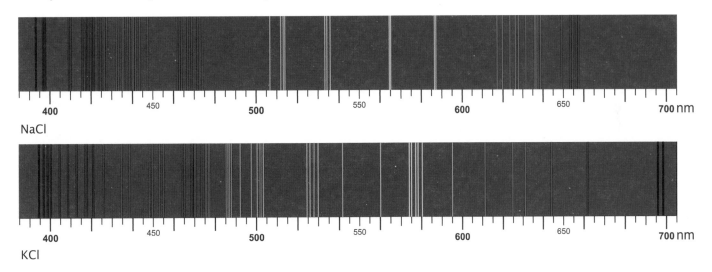

NaCl

KCl

The Group 2 elements are the Alkaline Earth Metals. These include Be (beryllium), Mg (magnesium), Ca (calcium), Sr (strontium), Ba (barium), and Ra (radium). $MgSO_4$ (magnesium sulfate) is Epsom salts that you use to soak swollen feet. $CaCO_3$ (calcium carbonate) is TUMS® that you may take as an antacid because it neutralizes excess stomach acid. Do the flame test with a small amount of Epsom salts and a small amount of ground up TUMS®. Are they similar?

The Group 17 elements are the Halogens. These include F (fluorine), Cl (chlorine), Br (bromine), I (iodine), and At (astatine). Look up and write a sentence for each, describing its properties. These elements can be quite toxic, so we are not going to do any procedures with them.

Write a report describing groups 1, 2, and 17. Describe the results of the flame tests and the answers to the questions. Write a paragraph describing the halogens.

CHAPTER 16

IONIC BONDS

OBJECTIVES AND VOCABULARY

The learning objectives of this lesson are to be able to demonstrate an understanding of:

1. The nature of ionic compounds

2. The difference between bond polarity and molecular polarity.

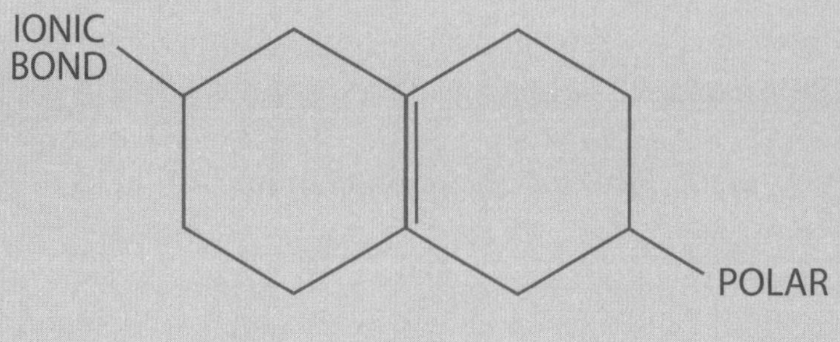

IONIC
BOND

POLAR

Na⁺ and Cl⁻ ions are attracted to each other because they have opposite charges. They do not exist as NaCl molecules but rather as crystals consisting of many layers where each layer consists of alternating Na⁺ and Cl⁻ ions. They form large crystals called sodium chloride or table salt. In the crystal, there are 6 Na⁺ ions around each Cl⁻ ion and 6 Cl⁻ ions around each Na⁺ ion. The attraction between the Na⁺ and Cl⁻ ions is called an **ionic bond**. It is a fairly strong attraction rather than an actual physical link. If you add NaCl to water, the NaCl readily dissolves, as is typical of most ionic compounds. Water molecules are **polar,** meaning that one end of the water molecule has a weak positive charge and the other end has a weak negative charge. Na⁺ ions are attracted to the negative end of water molecules and Cl⁻ ions are attracted to the positive end of water molecules. This pulls the NaCl crystal apart, dispersing them between the water molecules.

> **Ionic bonds *are a fairly strong attractions rather than actual physical links.***

The ionic bonds of NaCl can be easily explained by the electron configurations of Na and Cl.

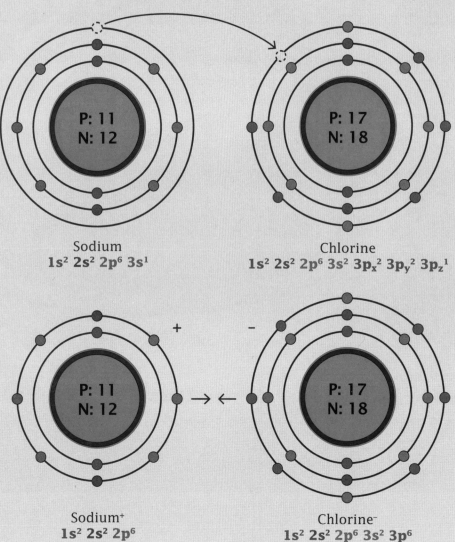

Sodium
$1s^2\ 2s^2\ 2p^6\ 3s^1$

Chlorine
$1s^2\ 2s^2\ 2p^6\ 3s^2\ 3p_x^2\ 3p_y^2\ 3p_z^1$

Sodium⁺
$1s^2\ 2s^2\ 2p^6$

Chlorine⁻
$1s^2\ 2s^2\ 2p^6\ 3s^2\ 3p^6$

Na has an atomic number of 11 giving it 11 positive charged protons. As a neutral atom, it also has 11 negative charged electrons with an electron configuration of …

$$1s^2\ 2s^2\ 2p^6\ 3s^1$$

This places the $1s^2$, $2s^2$, and $2p^6$ electrons closer to the nucleus. There is a fairly large energy gap between 2p and 3s orbitals. This makes it easier to remove the lone $3s^1$ electron. This is like saying that it is easier for an object in orbit around the Earth to leave the earth. After losing the $3s^1$ electron, the Na⁺ ion has the electron configuration …

$$1s^2\ 2s^2\ 2p^6$$

Cl has the atomic number of 17. As a neutral atom, a Cl atom has 17 electrons giving it an electron configuration of …

$$1s^2\ 2s^2\ 2p^6\ 3s^2\ 3p_x^2\ 3p_y^2\ 3p_z^1$$

The $3p_z^1$ orbital has 1 unpaired electron. It would be more stable if it were paired with another electron. As well, it is part of the 3p orbitals that already have 5 electrons. This means that the $3p_z$ electron is not isolated by itself. This makes it easier for Cl to take another electron to pair with the $3p_z^1$ electron rather than trying to lose all 5 of the 3p electrons. The Cl takes the $3s^1$ electron from the Na and places it into the $3p_z$ orbital of Cl. This gives the Cl⁻ ion 18 electrons with the electron configuration …

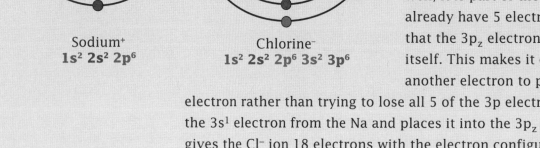

Sodium⁺
$1s^2\ 2s^2\ 2p^6$

Chlorine⁻
$1s^2\ 2s^2\ 2p^6\ 3s^2\ 3p^6$

$$1s^2\ 2s^2\ 2p^6\ 3s^2\ 3p^6$$

The Na$^+$ and Cl$^-$ ions satisfy the Octet Rule. The elements on the right side of the jagged line on the periodic table are non-metals and those to the left of the line are metals. A helpful reminder is that non-metals (like Cl) tend to gain electrons and metals (like Na) tend to lose electrons. This means that when non-metals come in contact with metals, the metal atoms give up a valence electron or electrons to the non-metal atoms. They become ions and form ionic bonds between them. This means that you can look at the periodic table and predict whether or not different elements will form crystals held together by ionic bonds. We are used to using the word salt to refer to NaCl because we use it on our food. But in chemistry, the word salt refers to any combination of a cation and a non-metal (or other anion) where they are held together by ionic bonds.

Consider F (fluorine). F has an atomic number of 9 with 9 electrons. The electron configuration of F is ...

$$1s^2 \, 2s^2 \, 2p_x^2 \, 2p_y^2 \, 2p_z^1$$

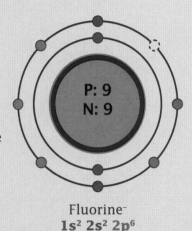

This gives F 7 electrons with the highest Principle Quantum Number of 2. It needs 1 more electron to complete the Octet Rule. By gaining 1 more electron its electron configuration becomes ...

$$1s^2 \, 2s^2 \, 2p^6$$

Fluorine$^-$
$1s^2 \, 2s^2 \, 2p^6$

F becomes F$^-$ by gaining 1 electron with 8 electrons now having a highest Principle Quantum Number of 2.

O (oxygen) has an atomic number of 8 with 8 electrons. Its electron configuration is ...

$$1s^2 \, 2s^2 \, 2p_x^2 \, 2p_y^1 \, 2p_z^1$$

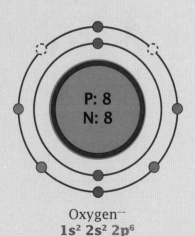

It has 6 electrons with the highest principle quantum number of 2. It needs 2 more electrons to give it 8 with the highest principle quantum number of 2. O gains 2 electrons becoming O^{--} with an electron configuration of ...

$$1s^2 \, 2s^2 \, 2p^6$$

Oxygen^{--}
$1s^2 \, 2s^2 \, 2p^6$

Consider Al (aluminum). Al has an atomic number of 13, giving it 13 electrons. Its electron configuration is ...

$$1s^2 \, 2s^2 \, 2p^6 \, 3s^2 \, 3p_x^1$$

This gives Al 3 electrons with the highest principle quantum number of 3. In this case, it is easier to lose 3 electrons than to gain 5 to complete

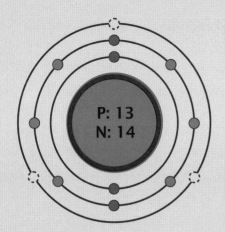

Aluminum^{+++}
$1s^2\ 2s^2\ 2p^6$

Hydrogen$^+$

the Octet Rule. Al loses 3 electrons becoming Al^{+++} with an electron configuration of ...

$$1s^2\ 2s^2\ 2p^6$$

H (hydrogen) with only 1 electron is more difficult. Usually when H is in contact with a non-metal, it gives up its electron. With Cl, it becomes H$^+$.

$$H + Cl \rightarrow H^+ + Cl^-$$

The H$^+$ is an acid and HCl is hydrochloric acid (the type formed in your stomach). H behaves differently when bonded with C (carbon). In this case it forms a covalent bond that is introduced in the next chapter.

To illustrate the reaction of an acid (H$^+$) with a metal, pour vinegar (acetic acid) to about ½ of an inch deep in a bowl. Place a penny in the bowl and come back and observe it about half an hour later. The small bubbles formed around the penny are hydrogen (H$_2$) gas, forming as the copper in the penny becomes brighter as the outer layer is being dissolved. The H$^+$ ions in the vinegar are gaining electrons from the Cu (copper) in the penny and the Cu is losing electrons as ...

$$2\ H^+ + 2\ e^-\ (electrons) \rightarrow H_2\ (g)$$

$$Cu \rightarrow Cu^{++}\ (aq) + 2e^-$$

When exposed to a metal the hydrogen is behaving like a non-metal in gaining electrons. Even though electrons are going from one atom to another, ionic bonds are not formed in this case because no negative ions are being formed. The Cu becomes positively charged and dissolves, but the H is going from a + charge to a neutral charge and escaping as a gas.

RELATIVE SIZES OF IONS

The radii of ions get larger as you go down the groups of the periodic table.

Ionic Radii

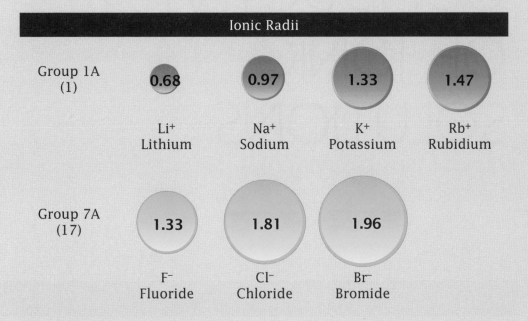

Group 1A (1)	0.68	0.97	1.33	1.47
	Li⁺ Lithium	Na⁺ Sodium	K⁺ Potassium	Rb⁺ Rubidium

Group 7A (17)	1.33	1.81	1.96
	F⁻ Fluoride	Cl⁻ Chloride	Br⁻ Bromide

This affects the sizes of the ionic crystals they form.

Ionic Salts

0.97 1.81	1.47 1.81	0.68 1.81
Na⁺Cl⁻	Rb⁺Cl⁻	Li⁺Cl⁻

CONDUCTIVITY OF IONIC SOLUTIONS

REQUIRED MATERIALS

- 3 beakers (100 ml)

- 1 beaker (250 ml)

- Stirring rod

- Laboratory scoop

- Scale

- Weighing boats

- Multimeter

- Table salt

- Sugar

- Baking soda

- 3 other common household liquids (vinegar, fruit juice, milk, etc.)

PURPOSE

This exercise is to demonstrate electrical conductivity as an important property of ionic compounds.

1. Prepare 5 percent, 10 percent, and 15 percent solutions of NaCl.

   A. Prepare 5 percent NaCl solution.

      i.   Weigh out 5 grams of NaCl.

      ii. Measure 95 ml of distilled water into the 250 ml beaker.

           Remember that ml (milliliters) of water can be used the same as grams of water. The 5 grams of NaCl and 95 grams of water add up to 100 grams.

      iii. Add the 5 grams of NaCl to the 95 ml of water and stir until the NaCl is completely dissolved.

      iv. Transfer 80 ml of the solution into a 100 ml beaker and dispose of the rest.

   B. Prepare 10 percent NaCl solution, following the same pattern as previously.

      i.   Weigh out _____ grams of NaCl.

      ii. Measure 90 ml of distilled water into the 250 ml beaker.

      iii. Add the NaCl to the water and stir until it is completely dissolved.

      iv. Transfer 80 ml of the solution into a 100 ml beaker and dispose of the rest.

   C. Prepare a 15 percent solution of NaCl.

      i.   Weigh out _____ grams of NaCl.

      ii. Measure _____ ml of distilled water into the 250 ml beaker.

      iii. Transfer 80 ml of the solution into a 100 ml beaker and dispose of the rest.

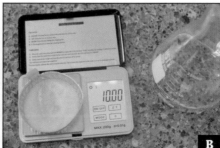

2. Test the conductivity (how well the solution conducts an electric current) of each of the 3 solutions.

A.

Using a multimeter on the resistance setting (Ω), set the dial to 2,000 ohms (Ω). This means that it will record up to 2,000 ohms of resistance. Turn the dial to this setting on the lower left of the meter. Place both leads into the 5 percent solution at opposite ends of the beaker, and record the number of ohms of resistance shown. The reading on the multimeter will probably start with a maximum resistance and then drop down and settle at the true value. The lower the resistance, the greater the conductivity. This is because pure water has a very high resistance to an electric current. As the Na^+ ions start migrating to the negative probe and the Cl^- ions migrate to the positive probe, the resistance will decrease and a current will flow between the probes.

B. Repeat Step A with the 10 percent solution.

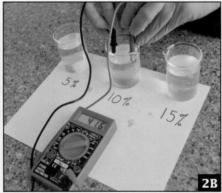
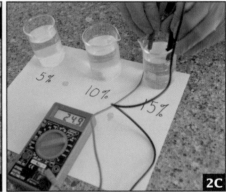

The conduction of an electric current through a solution means that electrons are leaving the negative electrode and going to the positive electrode. The lower the resistance, the better the solution is at conducting electricity. The negative electrode is negative because it has excess negatively charged electrons. The positive electrode is positively charged because it has more positively charged protons than negatively charged electrons. In the NaCl solution, the Na^+ and Cl^- ions come apart in water. The Na^+ ions are attracted to the negatively charged end of water molecules and the Cl^- ions are attracted to the positively charged ends of water molecules. When the Na^+ ions come in contact with the negative electrode of the multimeter, $Na^+ + e^- \rightarrow Na$. When Cl^- ions come in contact with the + electrode of the meter, electrons are given up from the Cl^- ions as
$2\ Cl^- \rightarrow Cl_2 + 2\ e^-$.

C. Repeat Step A with the 15 percent solution.

3. Prepare two separate 5 percent solutions of sugar (sucrose, table sugar) and sodium bicarbonate in 100 ml beakers.

4. Test and record the conductivity of the sucrose and sodium bicarbonate solutions.

5. Prepare samples of three other common household liquids (vinegar, fruit juice, milk, etc.) in 100 ml beakers.

6. Test and record the conductivity of these samples.

From your results, what can you conclude about each of these solutions? In your report, include how you made the NaCl, sucrose, and sodium bicarbonate solutions, your results, and whether or not each is an ionic compound.

CONDUCTORS OF ELECTRIC CURRENTS

LAB INSIGHTS

Ionic solutions are great conductors of electric currents. When Cl- ions donate electrons to + charged anodes (electrodes) and Na+ ions gain electrons from – charged cathodes, they complete the circuit between the two poles of the power source.

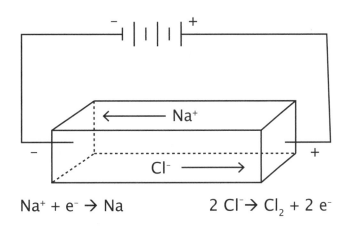

$$Na^+ + e^- \rightarrow Na \qquad 2\,Cl^- \rightarrow Cl_2 + 2\,e^-$$

CHAPTER 17

COVALENT BONDS

OBJECTIVES AND VOCABULARY

The Learning Objectives are for the students to gain an understanding (as evidenced by the students' performance on the quiz) of:

1. The nature of covalent bonds between non-metal atoms

2. Being able to predict which atoms will form covalent bonds

3. Being able to draw the Lewis structures of atoms covalently bonded together.

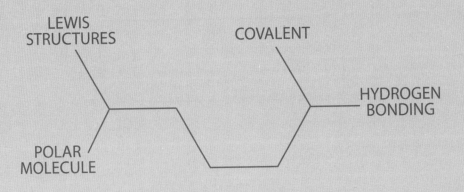

Ionic bonds form when metals combine with non-metals. What happens when non-metals combine with non-metals? When non-metals come in contact with non-metals they both want to acquire electrons. What happens when two little kids want the same toy? They are supposed to share. Likewise, non-metal atoms share outer electrons with each other. For each covalent (formed by a valence electron from each atom) bond, there is a pair of electrons shared between the two atoms. **Covalent** means that valence electrons belong to both atoms at the same time. Because the shared pair of electrons belong to both atoms, the atoms overlap in the same space. This forms an actual physical link rather than just an attraction.

A good example is the O_2 (oxygen molecule) that we inhale. The electron configuration of an O atom with 8 electrons is ...

$$1s^2\ 2s^2\ 2p_x^2\ 2p_y^1\ 2p_z^1$$

Covalent *means that valence electrons belong to both atoms at the same time.*

Oxygen
$1s^2\ 2s^2\ 2p^6$

Oxygen
$1s^2\ 2s^2\ 2p^6$

Each O atom needs to fill the $2p_y^1$ and $2p_z^1$ orbitals with 2 more electrons. When the O_2 molecule forms, the two O atoms share their $2p_y^1$ and $2p_z^1$ electrons with each other. This results in filling the $2p_y$ and $2p_z$ orbitals of both atoms. The overlapping $2p_y$ and $2p_z$ orbitals are called molecular orbitals. The two O atoms have 2 pairs of electrons shared between them so they have 2 covalent bonds. They are much stronger than ionic bonds and do not come apart in water. A good example is table sugar (a 12 carbon molecule called sucrose). The atoms in a sucrose molecule are held together by covalent bonds, so when you put sucrose into water, the sucrose molecules stay intact and taste sweet.

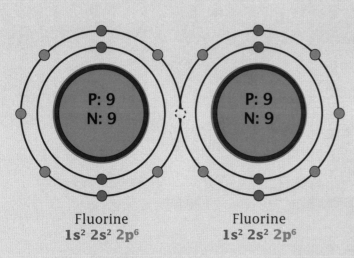

Sucrose

Another example is F_2, the natural form of fluorine gas. The electron configuration of F with 9 electrons is ...

$$1s^2\ 2s^2\ 2p_x^2\ 2p_y^2\ 2p_z^1$$

The $2p_z$ orbitals of each F atom are half filled. When 2 F atoms combine, they share their $2p_z^1$ electrons, producing a shared pair of electrons between them. This gives 2 F atoms with 1 shared pair of electrons between them, making 1 covalent bond. Remember that the O atoms have 2 pairs of electrons between them, so they have 2 covalent bonds. The 2 covalent bonds between the two O atoms are stronger than the 1 covalent bond between the F atoms.

Fluorine
$1s^2\ 2s^2\ 2p^6$

Fluorine
$1s^2\ 2s^2\ 2p^6$

The covalent bonds between the atoms are drawn as F–F and O=O.

N (nitrogen) atoms have 7 electrons with an electron configuration of ...

$$1s^2\ 2s^2\ 2p_x^1\ 2p_y^1\ 2p_z^1$$

There are 3 unpaired 2p electrons, which need 3 more electrons to fill their orbitals. N has 5 electrons with the highest principle quantum number of 2. It is easier to gain 3 electrons to make 8 for the Octet Rule than to lose 5 electrons. Therefore, 2 N atoms share 3 pairs of electrons between them, giving 3 covalent bonds. These 3 covalent bonds are

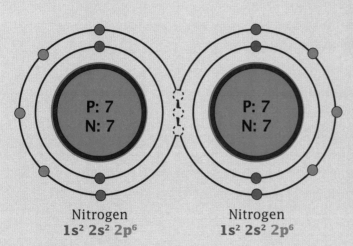

Nitrogen
$1s^2\ 2s^2\ 2p^6$

Nitrogen
$1s^2\ 2s^2\ 2p^6$

much stronger than the 2 covalent bonds between two O atoms or the 1 covalent bond between 2 F atoms. The N_2 molecule is drawn as

$$N \equiv N$$

Another way to indicate the covalent bonds is with **Lewis structures**. The electron dot symbols were original with G. N. Lewis while he was lecturing to a general chemistry class in 1902. As well, he contributed a lot to the early understanding of acids and bases. The outer shell electrons (those with the highest principle quantum number) are drawn around the symbol for each atom. F with 9 electrons and an electron configuration of $1s^2\ 2s^2\ 2p_x^2\ 2p_y^2\ 2p_z^1$ is drawn as ...

The 2 dots in between the F atoms represent the pair of electrons (covalent bond) between them. Notice that the 2s and 2p electrons (not the 1s electrons) are drawn for the F atoms because they have the highest principle quantum number (2). The 2 dots between the F atoms are the two $2p_z^1$ electrons shared between them and the other 6 dots around each F atom are the other unshared paired electrons ($2s^2\ 2p_x^2\ 2p_y^2$).

An O atom with 8 electrons has an electron configuration of $1s^2\ 2s^2\ 2p_x^2\ 2p_y^1\ 1p_z^1$. The O_2 molecule is drawn as ...

The 4 dots between the O atoms are the 2 pairs of shared electrons making 2 covalent bonds. They are formed by sharing their $2p_y^1$ and $2p_z^1$ electrons.

A N atom has 7 electrons with an electron configuration of $1s^2\ 2s^2\ 2p_x^1\ 2p_y^1\ 2p_z^1$. The Lewis structure of N is ...

The 3 single dots are the $2p_{x1}\ 2p_{y1}$ and $2p_{z1}$ electrons. The paired dots are the $2s^2$ electrons.

The N_2 molecule is drawn as ...

The 3 pairs of dots between the N atoms are the 3 shared pairs of electrons from the $2p_x^1$ $2p_y^1$ and $2p_z^1$ orbitals. This gives the N_2 molecule 3 covalent bonds. The 2 dots on the outside of each N atom are the $2s^2$ electrons.

A very unique and interesting molecule is the water molecule H_2O. A H atom with 1 electron ($1s^1$) is drawn as H. The Lewis electron dot drawing of a water molecule is ...

Bonding Electron Pairs

There is 1 covalent bond between the O atom and each of the H atoms. The O atom forms 2 covalent bonds in water (1 to each of the 2 H atoms).

Note: No matter which corners of the tetrahedron hydrogen is attached, the shape of the water molecule (tetrahedron) remains the same and the polarity remains the same.

Water is probably the most unique molecule in the universe. Its uniqueness accounts for its ability to support biological life. A water molecule has a tetrahedral shape that looks like a pyramid (with the oxygen atom in the middle) in which all of the 3 sides and bottom are identical and any point (top and 3 corners of the base) is the same distance from the others. Two of the points of the pyramid are the $2s^2$ and $2p_x^2$ electron pairs with their negative charges. The other 2 points of the pyramid are the electron pairs that bond the 2 H atoms. The protons of the H atoms at these 2 points give that side of the water molecule a positive charge. This gives a molecule that is negatively charged on one end and positively charged on the other end. Water is a **polar molecule**, meaning that it has opposite poles. The + and – poles are partial + and – charges so they are written as d (Greek letter delta) + and d -.

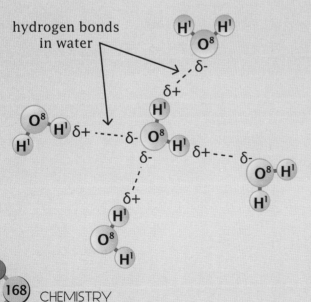

hydrogen bonds in water

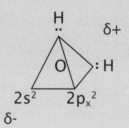

As a result there is an attraction between water molecules (the positive ends attracted to the negative ends) called **hydrogen bonding**. As in ionic bonds, these are not true bonds but attractions. Each attraction is very weak, but when you add them together, they are very substantial. Water molecules are very light, and if it were not for the hydrogen bonding, all water molecules would escape into the atmosphere and vaporize at temperatures lower than -200° F. This would not be good.

FROZEN WATER MOLECULES

Hydrogen bonding between water molecules in the frozen state causes them to be farther apart in a hexagonal (six-sided) structure so that there are fewer water molecules in the same volume of ice than in the same volume of liquid water. This lowers the density of ice so that it floats. This is also the arrangement of water molecules in a snowflake.

LAB INSIGHTS

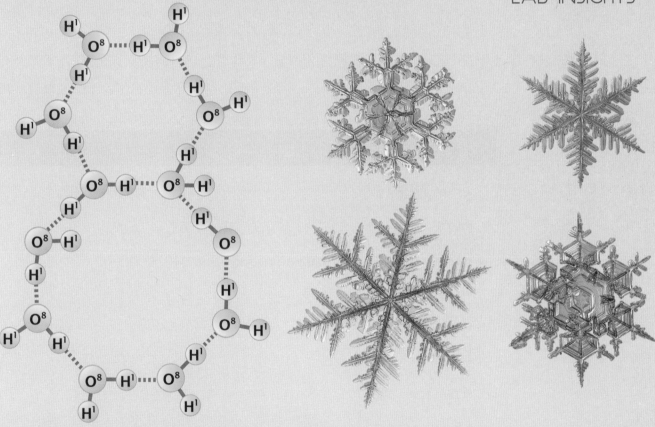

If ice did not float in the lakes and rivers in northern latitudes, the fish and other life forms would die because the water would freeze at the bottom. This is God's grace coming through in design and beauty.

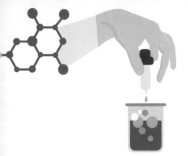

POLAR AND NON-POLAR MOLECULES

REQUIRED MATERIALS

- Petri dish

- Milk

- Grease pencil or permanent marker

- Toothpick

- Dish soap

- Food coloring (4 colors)

- Test tube

- Stirring rod

- Olive oil

- Table salt

- Laboratory scoop

PURPOSE

This exercise is designed to demonstrate the nature of polar and non-polar molecules and how to distinguish them from each other.

1. Turn a Petri dish over and mark 4 pie-shaped sections on the underside and sides of the dish.

2. Pour milk into the upright marked Petri dish to a depth of about 1 cm (centimeter). Place a couple of drops of food coloring into the milk with a different color in each of the four areas around the outer edges of the Petri dish.

3. Dip a toothpick into liquid dish soap and place a drop of the dish soap in the milk in the center of the Petri dish.

4. Describe what happens.

   The dish soap molecules have one end that is polar with the other end non-polar. The polar end of the dish soap molecule attracts polar molecules and the non-polar end of the dish soap attracts non-polar molecules. With this in mind, explain what you observed in the Petri dish.

5. Fill a test tube 1/3 full with water.

6. Add an equal amount of olive oil to the test tube and allow it to settle.

7. Mix the liquid in the tube with a glass stirring rod and allow it to settle. Which liquid is on top?

8. Add some NaCl to the tube and observe what happens to it. What does that say about the polarity of NaCl?

9. Add a drop of liquid dish soap and stir the mixture. What happens? What does this tell you about the dish soap?

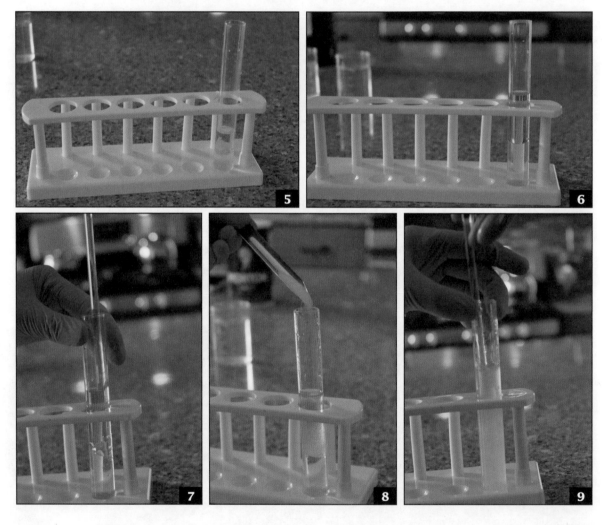

Prepare your report and describe your procedures and results. Answer the questions.

HYDROPHOBIC AND HYDROPHILIC

LAB INSIGHTS

In the lab exercise, you used dish soap to disperse oil molecules. Fatty acid molecules have a hydrophobic (water fearing) end that dissolves in oil and a hydrophilic (water loving) end that dissolves in water.

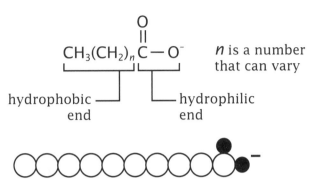

n is a number that can vary

Soap consists of fatty acids. The hydrophobic end dissolves in oil and the hydrophilic end dissolves in water. Soap disperses oil into tiny droplets that can be washed away by water — without which we could not bathe, wash dishes, or do laundry. God has abundantly provided for us in so many ways. This is what we take for granted but should not. We need to thank God for His abundant grace and blessings.

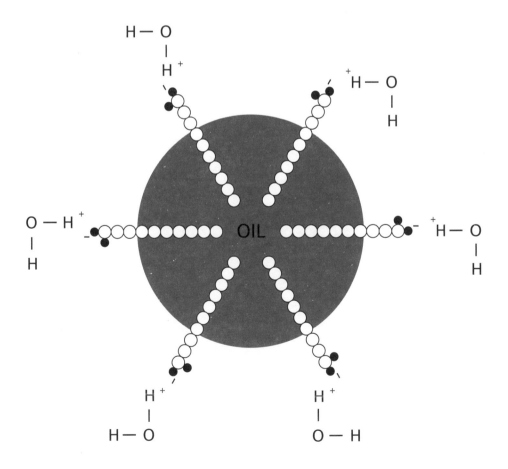

CHAPTER 18

METAL ATOMS

OBJECTIVES AND VOCABULARY

The Learning Objectives are for the students to gain an understanding (as evidenced by their performance on the quiz) of:

1. The properties of metals and electric currents

2. How the properties of metals are explained by the arrangements of their atoms and outer electrons

3. The properties of Alkali Metals, Alkaline Earth Metals, transition metals and metalloids

4. The oxidation and reduction of metals.

DUCTILE   ALTERNATING CURRENT

CATIONS   REDUCTION

DIRECT CURRENT   ANIONS

OXIDIZER

OXIDATION   VOLTAGE

MALLEABLE   REDUCER

With ionic bonding, electrons are transferred from metal atoms to non-metal atoms. In covalent bonds, pairs of electrons are shared between non-metal atoms. But what about when metal atoms are around other metal atoms where they all want to get rid of outer electrons. Where can the electrons go without a non-metal to take them? God designed metal atoms (about 80 percent of the known elements) so that some of their outer electrons can move about between the atoms.

Electrons between metal atoms have greater mobility because they are not tied down to any one atom.

Metals are good conductors of electricity, non-metals are not. Electrons between metal atoms have greater mobility because they are not tied down to any one atom. You have certainly noticed how quickly an electric current (movement of charged objects like electrons) appears to move through a wire. You turn on a light switch and immediately the light comes on. Even though it seems like it, the electrons do not actually travel all the way through the wire that fast. The electrons that enter the wire on one end are not the same electrons that leave the wire on the other end. Picture the wire as if it were a long narrow glass tube filled single file with marbles. If you poke a marble into one end of the tube, one falls out of the other end. Electric currents flow through a wire in a similar fashion. In a **direct current (DC)**, using the analogy, as marbles are poked in one end of the tube, they keep falling out the other end. Electrons entering the wire on one end cause other electrons to come off the other end of the wire.

In an **alternating current (AC)**, a marble is pushed in one end of the tube and one falls out the other end. Then one is pushed back in at the opposite end causing the first marble to fall out. The process is continually repeated back and forth. In an electric current, an electron enters one end of the wire causing one to leave the other end. Then it is reversed, and the electron that left the other end goes back into the wire and the first electron leaves the wire. This is a very efficient process but is only possible because the outer electrons of the metal atoms can move between the metal atoms.

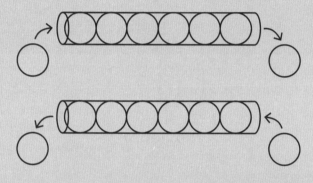

Besides being good conductors of electricity, metals are good conductors of heat. The metal atoms have the opportunity to move around faster in solid metals as they "float" around in a "sea of electrons". The temperature of a substance is a measure of how fast its atoms or molecules are moving. Metal atoms have more freedom to move about faster allowing their temperature to rise. This is why you do not want to grab a metal handle of a pan that has been heating on the stove.

Metals are also **malleable,** meaning that they can be pounded out into flat sheets. It is easier for layers of metal atoms to slide past each other because they are separated by the loose electrons between them. For the same reason, they are **ductile,** meaning that they can be pulled out into narrow wires.

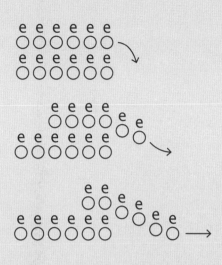

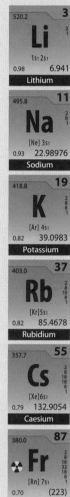

520.2			3
	Li		2
			1
	1s1 2s1		
0.98		6.941	
Lithium			

495.8			11
	Na		2
			8
			1
	[Ne] 3s1		
0.93		22.98976	
Sodium			

418.8			19
	K		2
			8
			8
			1
	[Ar] 4s1		
0.82		39.0983	
Potassium			

403.0			37
	Rb		2
			8
			18
			8
			1
	[Kr]5s1		
0.82		85.4678	
Rubidium			

357.7			55
	Cs		2
			8
			18
			18
			8
			1
	[Xe]6s1		
0.79		132.9054	
Caesium			

380.0			87
❖	Fr		2
			8
			18
			32
			18
			8
			1
	[Rn] 7s1		
0.70		(223)	
Francium			

Metals usually have a shiny "metallic" luster. This is because electrons moving between the metal atoms are said to be in energy bands, which are similar to molecular orbitals of covalent compounds. When light shines onto a metal surface, the electromagnetic energy (light) is absorbed and given back off by the electrons in the energy band, resulting in a metallic luster.

On the periodic table, the elements in the lower left corner have the strongest metallic properties and those in the upper right corner display these properties the least (being non-metals). Some metals (in the lower left corner of the periodic table) release their outer electrons more readily than others. This difference in the ability to hold onto and release outer electrons results in **oxidation** (giving up electrons) and **reduction** (taking up electrons). This difference between different metals is the basis for the function of most batteries.

As we looked at before, the metals in Group 1 (Li, Na, K, Rb, Cs and Fr) are the Alkali Metals. They all have s^1 electrons that are readily given up. This makes them very reactive metals. When they come in contact with water, they react violently, releasing tremendous amounts of energy. We do not normally think of metals as bursting into flame or exploding when coming in contact with water. When Na comes in contact with water, water molecules come apart as OH$^-$ (called a base or alkali) and H$^+$ acid. The 3s^1 electron from Na goes to the H$^+$ ions, forming H$_2$ (hydrogen gas).

$$2 \text{ Na} \rightarrow 2 \text{ Na}^+ + 2 \text{ e}^- \text{ (electron)}$$
$$2 \text{ H}_2\text{O} \rightarrow 2 \text{ OH}^- + 2 \text{ H}^+$$
$$2 \text{ H}^+ + 2 \text{ e}^- \rightarrow \text{H}_2$$

Sodium metal

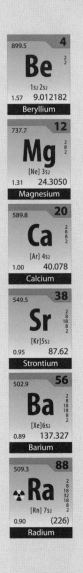

Notice that the Na is oxidized (loses electrons) and the H⁺ is reduced (gains electrons). An alkaline (OH⁻) solution is produced, so the metal Na is called an alkali metal.

The metals of Group 2 of the periodic table are the Alkaline Earth metals. They include Be, Mg, Ca, Sr, Ba, and Ra. Just like the alkali metals, these elements are not found naturally by themselves because they are very reactive giving up their s^2 electrons. As an example, Mg (magnesium) reacts with O_2 (oxygen) to produce MgO (magnesium oxide).

$$2\ Mg + O_2 \rightarrow 2\ MgO$$

This can be understood with the following:

$$2\ Mg \rightarrow 2\ Mg^{++} + 4\ e^-$$
$$O_2 + 4\ e^- \rightarrow 2\ O^{--}$$
$$2\ Mg^{++} + 2\ O^{--} \rightarrow 2\ MgO$$

The Mg gives its $3s^2$ electrons to an O atom. The Mg is oxidized and O is reduced by the Mg.

The alkaline earth metals are often naturally found as hydroxides (combined with OH⁻) such as $Mg(OH)_2$. The released OH⁻ gives this group the name alkaline earths. Mg is the second most abundant metal found in the sea (Ca is the most abundant) and is also found in association with SiO_2 (silicon dioxide) in sand.

All of the metals between Groups 2 and 13 are the transition metals. Their properties are better understood by looking at their partially filled

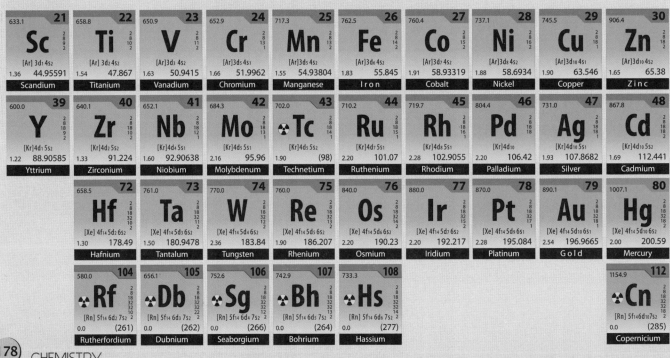

d and f orbitals. They are heavier metals and are usually found naturally in alloys (combined with other metals) or by themselves.

Copper pipes

Several elements along the line separating the metals from the non-metals are called metalloids or semiconductors. They got the name semiconductor because they do not conduct electricity as well as other metals. At the same time, they do not accept electrons as well as non-metals. They are very useful in integrated circuits and solid state electronics. This is where different pathways for the flow of electric currents can be set up in crystals rather than with a maze of wires. This has made the circuitry possible in miniature form in computers and other electronic devices. Among the most significant are Si and Ge.

To review, positive charged ions, like Mg^{++} and Na^+, are called **cations**. Negative ions formed by non-metals, like Cl^- and OH^-, are called **anions**.

Protons	1	1	1
Electrons	0	1	2
Charge	+1	0	-1
Notation	H^+	H	H^-
Classification	cation	neutral (not an ion)	anion

Plumbers say to never attach galvanized pipe to copper pipe. Galvanized pipe is covered with Zn (zinc). When Zn and Cu (copper) come together, a battery is produced. Zn gives up its outer 2 electrons more readily than Cu gives up its electrons. As metals, they both want to give away the outer electrons, but Zn can get rid of the electrons easier than Cu. The movement of electrons from Zn to Cu is an electric current. The force with which the Zn forces its electrons onto the Cu is measured as **voltage**.

$$Zn \rightarrow Zn^{++} + 2\ e^-$$
$$Cu^{++} + 2\ e^- \rightarrow Cu$$

The battery contains Zn solid metal, Zn^{++} in solution, solid Cu, and Cu^{++} in solution.

Cu is an **oxidizer** because it oxidizes (removes electrons) from Zn. Zn is a **reducer** because it gives electrons to the Cu^{++}. It would be interesting if a zinc-covered pipe were connected to a copper pipe used to carry gas. Boom! This is why black iron (not coated with zinc) pipe is used in gas lines. Chemistry can be very practical.

LABORATORY 18

OXIDATION-REDUCTION

REQUIRED MATERIALS

- Zinc metal
- Copper metal
- Lemon
- Iron nails
- Sandpaper
- Vinegar
- Petri dish
- Multimeter
- NaCl solution

INTRODUCTION

In earlier lessons, it was stated that metal atoms tend to give up valence electrons. That is true, but some metals tend to give up electrons more forcefully than others. When Zn (zinc) and Cu (copper) are brought close together, each Zn atom gives up 2 electrons and the Cu ion (Cu^{++}) is forced to receive them.

$$Zn\ (s) \rightarrow Zn^{++}\ (aq) + 2\ e^-$$
(s) means solid and (aq) means aqueous or dissolved in water

$$Cu^{++}\ (aq) + 2\ e^- \rightarrow Cu\ (s)$$

PURPOSE

Upon the completion of this exercise the student should have a clearer understanding of the oxidation and reduction of metals and ions.

PROCEDURE

1. Observe oxidation and reduction reactions:

   A. Take a couple of iron nails and rub them with a piece of sand paper to remove any coating that they may have on them.

   B. Place them in a container of water and leave them there for several days so that they become rusty.

      In this case, the iron reacts with oxygen. The product is iron oxide (rust). Which is oxidized, the iron or the oxygen? Which is reduced, the iron or the oxygen?

   C. Afterward, place the rusty nails in vinegar overnight.

      What does the acetic acid in the vinegar do to the rusty nails? Is the rust being oxidized or reduced?

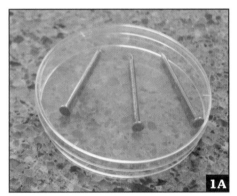

1A

1B

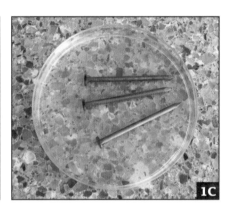

1C

2. Create a lemon battery:

   A. Take a piece of zinc metal and a piece of copper metal.

      If necessary, cut a piece from each small enough to poke into a lemon without touching (several inches long, and less than an inch wide).

   B. Insert the zinc and copper about an inch into a fresh lemon, placing them so that they are parallel about ½ inch apart. It is important that they not touch each other inside the lemon.

      When the electrons move from the zinc to the copper, the electrons are carried through the lemon juice by the acid, just like they were carried by the Na^+ and Cl^- ions in an earlier lab exercise.

2B

2C

C. Test the voltage of your lemon battery.

   i.   Use a multimeter by placing the leads on the copper and zinc strips, sticking out of the lemon using the (V) range.

   ii.   Another test is to place the copper and zinc ends into a NaCl solution salty enough for you to taste it.

   If small bubbles appear in the salt water, your battery is working. The following reaction is occurring in the salt water:

$$2 \; Na^+ + 2 \; e^- \rightarrow 2 \; Na \text{ and } 2 \; Cl^- \rightarrow Cl_2 + 2 \; e^-$$

   Even though sodium metal is explosive and chlorine gas is poisonous, you will not produce enough to be concerned about.

   iii.   If you do not have enough current to produce noticeable bubbles, place the ends of the copper and zinc on your tongue and see if it tingles. If it does, your lemon battery is working. It does not produce enough voltage or current to worry about.

LAB INSIGHTS

RUSTING

The corrosion (rusting) of iron (Fe) is a redox (reduction – oxidation) reaction. It has been estimated that 20 percent of the iron products produced annually in the United States is to replace things that rusted and were discarded.

Fe is oxidized by O_2.

$$2 \; Fe \rightarrow 2 \; Fe^{++} + 4 \; e^-$$
$$O_2 + 4 \; H^+ + 4 \; e^- \rightarrow 2 \; H_2O$$
$$4 \; Fe^{++} + O_2 + 4 \; H_2O \rightarrow 2 \; Fe_2O_3 + 8 \; H^+$$

Fe_2O_3 is rust.

SINGLE REACTIONS

When zinc (Zn) is placed in acid (such as citric acid in lemon juice) it gives up electrons to the acid forming zinc ions (Zn^{++}) and hydrogen gas (H_2).

$$Zn \rightarrow Zn^{++} + 2\ e^-$$

$$2\ H^+\ (acid) + 2\ e^- \rightarrow H_2$$

$$Zn + 2\ H^+ \rightarrow Zn^{++} + H_2$$

The bubbles that form are H_2 gas. Copper placed in lemon juice is also oxidized (loses electrons).

$$Cu \rightarrow Cu^{++} + 2\ e^-$$

$$2\ H^+ + 2\ e^- \rightarrow H_2$$

$$Cu + 2\ H^+ \rightarrow Cu^{++} + H_2$$

The zinc (Zn) and copper (Cu^{++}) ions react where the Zn loses 2 electrons (oxidized) and the Cu^{++} ions gain 2 electrons (reduced). Metals do not readily accept electrons. The zinc releases the electrons stronger than copper, so the copper ions are forced to take the electrons from the zinc, producing an electric current.

$$Zn \rightarrow Zn^{++} + 2\ e^-$$

$$Cu^{++} + 2\ e^- \rightarrow Cu\ (solid\ metal)$$

$$Zn + Cu^{++} \rightarrow Zn^{++} + Cu$$

LAB INSIGHTS

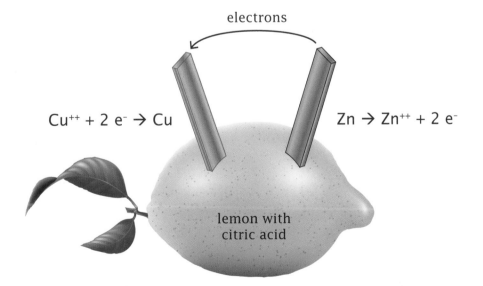

electrons

$Cu^{++} + 2\ e^- \rightarrow Cu$ $Zn \rightarrow Zn^{++} + 2\ e^-$

lemon with
citric acid

BATTERIES

OBJECTIVES AND VOCABULARY

The Learning Objectives are for the students to gain an understanding (as evidenced by the students' performance on the quiz) of:

1. The principles involved in the function of a battery

2. The function of an alkaline battery

3. The function of a 12V car battery

4. The function and uses for a hydrogen fuel cell as a source of electric power.

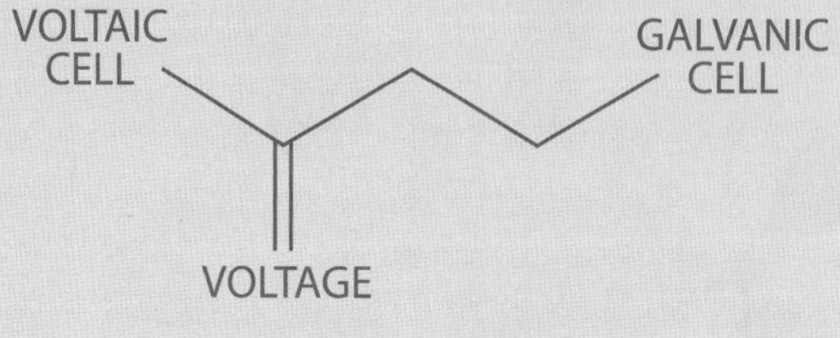

VOLTAIC
CELL   GALVANIC
CELL

VOLTAGE

We have come to depend upon batteries until we cannot live without them. We carry portable power around in our pocket.

A battery begins with a **voltaic cell** (also called a **galvanic cell**). It consists of oxidation — reduction reactions in which electrons move through an external circuit from atoms being oxidized to those being reduced.

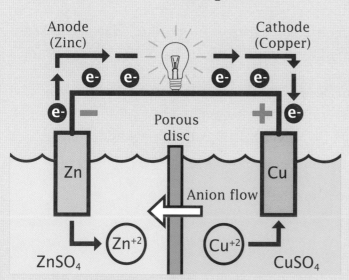

The anode is the negative post of a battery, and the cathode is the positive post of a battery. The anode is the source of the electrons (where oxidation occurs) and the cathode is where the electrons go (where reduction occurs). The anode attracts the anions (which want electrons) and the cathode attracts the cations (which easily lose electrons). The negatively charged electrons from the anode are attracted to the positive charges of the protons of the atoms to be reduced.

The **voltage** or EMF (electromotive force) of a battery is also called the standard cell potential.

The word standard (meaning at room temperature) is indicated by the superscript °. The E^o cell represents the force moving the electrons in the battery. The moving of the electrons is the electric current produced by the battery. There has to be 2 reactions, one giving the electrons away and the other being forced to take the electrons. All metals tend to lose electrons, so the total force moving the electrons in the battery is that of the metal forcing the electrons on the other metal minus the force of the other metal trying to resist taking the electrons. Think of it as someone forcing something onto someone that does not want it.

A typical alkaline battery consists of the reactions ...

Anode (oxidation – losing electrons)

$$Zn \ (s) + 2 \ OH^- \ (aq) \rightarrow Zn(OH)_2 \ (s) + 2 \ e^- \ (electrons)$$

The zinc atoms are oxidized $Zn \rightarrow Zn^{++}$

In $Zn(OH)_2$ the zinc atom has a +2 charge.

Cathode (reduction – gaining electrons)

$$2 \ MnO_2 \ (s) + 2 \ H_2O \ (l) + 2 \ e^- \rightarrow 2 \ MnO(OH) \ (s) + 2 \ OH^- \ (aq)$$

The manganese atoms are reduced $Mn^{+4} + 2 \ e^- \rightarrow Mn^{++}$

They are called alkaline batteries because the OH^- from KOH (potassium hydroxide) is alkaline.

A common alkaline battery purchased today has an EMF (electromotive force) of 1.55 volts at room temperature.

A 12 volt lead-acid car battery is made up of 6 voltaic cells in series (connected end to end with the + poles touching the – poles). Their voltages are added to each other.

The Anode (oxidation) is ...

$$Pb \ (s) + HSO_4^- \ (aq) \rightarrow PbSO_4 \ (s) + H^+ \ (aq) + 2 \ e^-$$

Electron flow

Paste of manganese dioxide and carbon

Fabric separator

Cathode (Graphite rod)

Anode (Zinc inner case)

Load

A typical alkaline battery

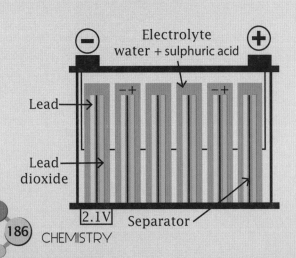

Electrolyte water + sulphuric acid

Lead

Lead dioxide

2.1V   Separator

12V lead-acid battery

Pb is lead. HSO_4^- is formed from sulfuric acid losing H^+ and is an acid.

The Cathode (reduction) is ...

$$PbO_2 \text{ (s)} + HSO_4^- \text{ (aq)} + 3 \text{ } H^+ \text{ (aq)} + 2 \text{ } e^- \rightarrow PbSO_4 \text{ (s)} + 2 \text{ } H_2O \text{ (l)}$$

The parts of the voltaic cell are Pb (lead) and H_2SO_4 (sulfuric acid).

Each voltaic cell has an EMF of 2 V (volts). Together the 6 voltaic cells in series provide 6 x 2 = 12 V.

An electric current can also be produced by a hydrogen fuel cell. They have been used by NASA as electrical energy sources for space craft. The by products are electricity and water, which can be drunk by the astronauts. One major drawback is that liquid oxygen and liquid hydrogen have to be kept at extremely low temperatures.

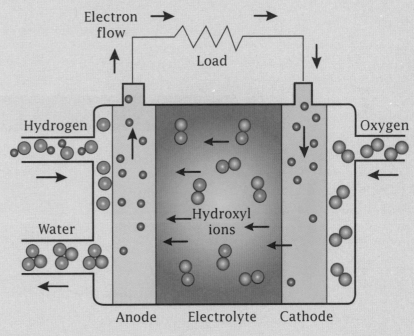

Hydrogen fuel cell

For a hydrogen fuel cell the cathode reaction is ...

$$O_2 \text{ (g)} + 2 \text{ } H_2O \text{ (l)} + 4 \text{ } e^- \rightarrow 4 \text{ } OH^- \text{ (aq)}$$

And the anode reaction is ...

$$2 \text{ } H_2 \text{ (g)} + 4 \text{ } OH^- \text{ (aq)} \rightarrow 4 \text{ } H_2O \text{ (l)} + 4 \text{ } e^-$$

The total reaction is ...

$$2 \text{ } H_2 \text{ (g)} + O_2 \text{ (g)} \rightarrow 2 \text{ } H_2O$$

The electric current is produced as the electrons flow from the anode to the cathode. The E^o cell = 1.23V.

Greater voltage is produced by connecting many hydrogen fuel cells in series (end to end).

A fuel cell is technically not a battery because it requires fuel and is not self-contained. Batteries for electric cars are heavy, expensive, and do not last very long before needing to be recharged even though progress has been made in improving their performance and cost. Fuel cells are being researched for automobiles and small portable devices. Instead of recharging or changing batteries, the fuel flows through a cell as needed. Of course, you would not be using liquid oxygen and liquid hydrogen to power your cell phone. It would be too cold and explosive.

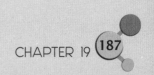

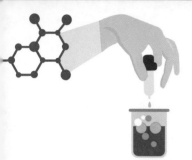

Zn/Cu GALVANIC CELL

REQUIRED MATERIALS

- Beaker (either size)

- Zinc strip

- Copper strip

- Vinegar

- Wire with alligator clips

- Multimeter

PURPOSE

This exercise applies what was learned in exercise #18 to the concept of a working battery.

PROCEDURE

1. Place a piece of Zn metal and a piece of Cu metal in a beaker about half full of vinegar (acetic acid). Leave it like this until small bubbles start to appear at one of the metal strips.

   This is similar to the lemon battery, except that the lemon has citric acid and this set-up has acetic acid.

2. Place the metal strips about an inch apart and connect them with a wire with clamps.

   Do you notice any bubbles forming around one of the metal strips? That one is being oxidized because it is giving electrons to the H+ ions from the acetic acid.

$$2 \ H^+ + 2 \ e^- \rightarrow H_2 \ (gas)$$

   At which metal is this occurring? Rewrite this reaction substituting the metal with Zn or Cu, depending upon which one is being oxidized (reducing the H^+).

$$Metal \rightarrow Metal^{++} + 2 \ e^-$$

3. Use a multimeter to test the voltage of your battery.

Do some research and look up how a AA battery is constructed. Draw a diagram of the inside of the battery and explain what is being reduced and what is being oxidized.

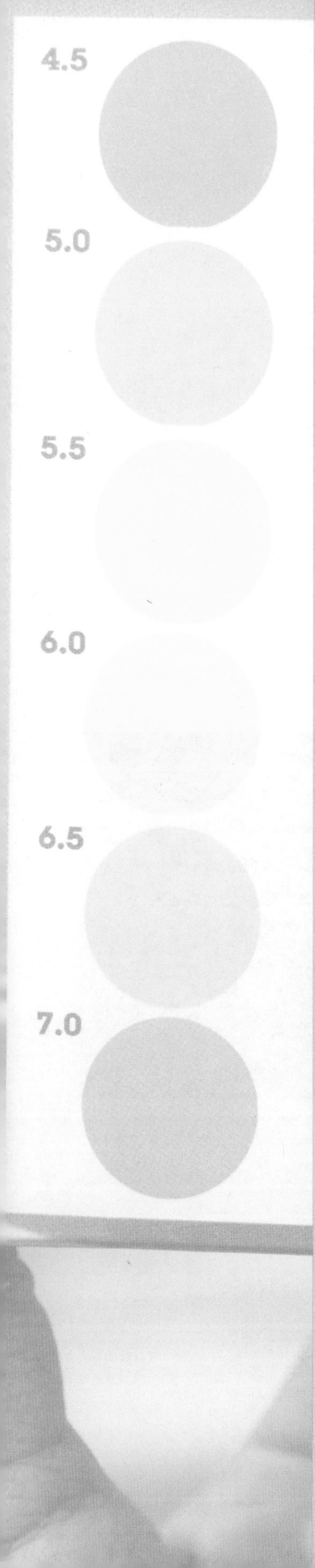

CHAPTER 20

ACIDS AND BASES I

4.5

5.0

5.5

6.0

6.5

7.0

OBJECTIVES AND VOCABULARY

The Learning Objectives are for the students to gain an understanding (as evidenced by their performance on the quiz) of:

1. Acids and bases

2. Strong acids and weak acids

3. Strong bases and weak bases

4. The pH scale

5. How to find the pH from $[H^+]$.

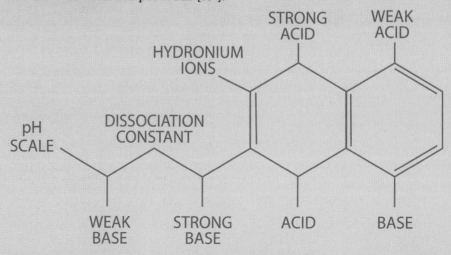

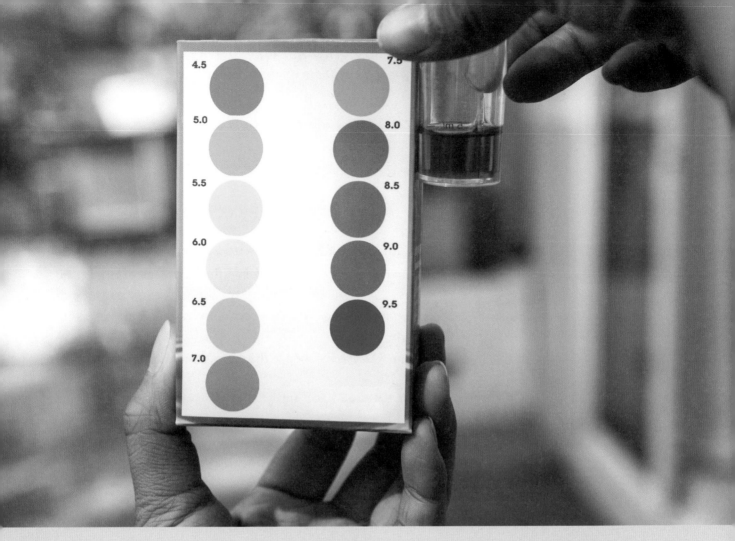

A H⁺ ion is an acid and a OH⁻ is a base. You are probably quite aware of how reactive they can be from life experiences. It may not be obvious but they are always present in water solutions. In water, H⁺ combines with H_2O to form H_3O^+ ions called **hydronium ions**. It is usually written as H⁺ as a shorter form, even though we know that it is really H_3O^+.

In distilled water with nothing else added to it, there are always as many hydronium ions as hydroxide (OH⁻) ions. This is because 1 hydronium ion and 1 hydroxide ion are being formed and recombined to form water molecules many times a second.

$$H_2O \quad + \quad H_2O \quad \longleftrightarrow \quad H_3O^+ \quad + \quad OH^-$$

If there is another source of H⁺ besides water, there will be more H⁺ than OH⁻ and the solution is called an **acid** and is said to be acidic. Some molecules, like HCl (hydrochloric acid) found in toilet bowl cleaner and stomach acid release all of their H^+ ions and are called **strong acids**.

$$HCl \rightarrow H^+ + Cl^-$$

Some molecules like acetic acid found in vinegar release only some of their H^+ ions and are called **weak acids**.

$$CH_3COOH \longleftrightarrow CH_3COO^- + H^+$$

Whether an acid is strong or weak depends upon whether or not they release all of their H^+ ions – not whether they are dilute or concentrated.

These terms can be a bit tricky. An acid, like HCl, can be a strong acid but still very dilute and release fewer H^+ ions (because it is dilute) than a weak acid that is more concentrated. Whether an acid is strong or weak depends upon whether or not they release all of their H^+ ions — not whether they are dilute or concentrated.

If there is another source of OH^- ions besides water, there will be more OH^- ions than H^+ ions and the solution is called a **base** and is basic. Some molecules, like NaOH (sodium hydroxide) found in drain cleaner, release all of their OH^- ions and are called **strong bases**.

$$NaOH \rightarrow Na^+ + OH^-$$

Some molecules, like NH_4OH (ammonium hydroxide) found in some window cleaners, release only a few of their OH^- ions and are called **weak bases**.

$$NH_4OH \longleftrightarrow NH_4^+ + OH^-$$

H^+	pH Scale	H^+
10^0 mol/l	0	1
10^{-1} mol/l	1	0.1
10^{-2} mol/l	2	0.01
10^{-3} mol/l	3	0.001
10^{-4} mol/l	4	0.0001
10^{-5} mol/l	5	0.00001
10^{-6} mol/l	6	0.000001
10^{-7} mol/l	7	0.0000001
10^{-8} mol/l	8	0.00000001
10^{-9} mol/l	9	0.000000001
10^{-10} mol/l	10	0.0000000001
10^{-11} mol/l	11	0.00000000001
10^{-12} mol/l	12	0.000000000001
10^{-13} mol/l	13	0.0000000000001
10^{-14} mol/l	14	0.00000000000001

A scale called the **pH scale** indicates the amount of acid or base in a water solution. Because water molecules give up both H^+ and OH^- ions, some of each are always present in water. The pH scale goes from 0 to 14 with 7 being in the middle. When the pH is 7.0, $[H^+]$ = $[OH^-]$. The square brackets [ ] stand for molarity or moles per liter. You might need to review this from an earlier chapter. Therefore, a solution of pH = 7.0 is neither acidic nor basic, even though some acid and base are present.

The pH scale shows increasing acid as the numbers go below 7 and increasing base as the numbers go above 7. A solution of pH = 6.0 has 10 x more $[H^+]$ than a solution of pH 7.0. A solution of pH 5.0 has 100 x more $[H^+]$ than a solution of pH 7.0. This is like the

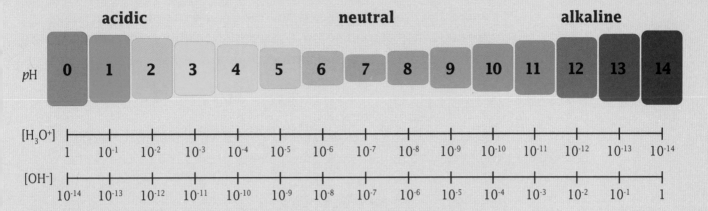

Richter Scale that is used to measure the strength of earthquakes. It is divided by 10s. A solution of pH 4.0 has 1,000 times more [H+] than pH 7.0.

The same pattern works for pH values greater than 7. A solution of pH 8.0 has 10 x more [OH−] than a solution of pH 7.0. A solution of pH 9.0 has 100 x more [OH−] than a solution of pH 7.0. As well, a solution of pH 8.0 has 1/10 as much [H+] as a solution of pH 7.0.

As the pH goes from 7 to 5, the [H+] increases by 100 or 10^2. As the pH goes from 7 to 4, the [H+] increases by 1,000 or 10^3. Do you see the pattern here? 7 − 5 = 2 and the [H+] increased by 10^2. 7 − 4 = 3 and the [H+] increased by 10^3?

This can also be seen for the [OH−]. As the pH goes from 7.0 to 9.0, the [OH−] increases 100 or 10^2 times. 9 − 7 = 2. As the pH goes from 7.0 to 10.0, the difference between 7 and 10 is 3 and the [OH−] increases by 1,000 or 10^3.

Another interesting pattern can be seen here as well. As the [H+] goes up, the [OH−] goes down. And as the [OH−] goes up, the [H+] goes down. This is because extra H+ ions combine with OH− ions to form H_2O. And extra OH− ions combine with H+ ions to form H_2O. There is an equilibrium or balance between the H+ ions and the OH− ions. This makes their behaviors very predictable. This is very important because a major goal of chemistry is to be able to predict what will happen under particular circumstances. As well, it gives testimony to our wise Creator who demonstrates His wisdom through these patterns. As well, it is a show of His grace because without these patterns biological life would be impossible. This accounts for the solubilities of the electrolytes in our body fluids that enables our enzymes and other proteins to function properly.

There is an equilibrium or balance between the H+ ions and the OH− ions.

The predictability of the H⁺ ions and the OH⁻ ions can be expressed numerically. The product of their concentrations is a constant at constant temperature called the **dissociation constant**. It is expressed as …

$$K_W = [H^+][OH^-] = 1.0 \times 10^{-14}$$

The "w" in the expression stands for water. Remember that for water, 1 out of every 10 million molecules come apart to form H⁺ and OH⁻ ions. 10 million is 10^7. For a pH of 7.0, $[H^+] = 10^{-7}$ and $[OH^-] = 10^{-7}$. 10^{-7} is 1 out of 10 million. This makes the K_W for water …

$$K_W = [1.0 \times 10^{-7}][1.0 \times 10^{-7}] = 1.0 \times 10^{-14}$$

Notice the numbers 10^{-7} for $[H^+]$ and $[OH^-]$ and 10^{-14} for K_W. This is where the number 7 for neutral pH comes from. The exponent of 10 is called a logarithm. The pH is the negative of the logarithm of the $[H^+]$.

If $[H^+]$ is 1.0×10^{-5}, the pH is 5.0. What will the $[OH^-]$ be? With the dissociation constant, we get …

$$[H^+] \, [OH^-] = [1.0 \times 10^{-5}] \, [OH^-] = 1.0 \times 10^{-14}$$

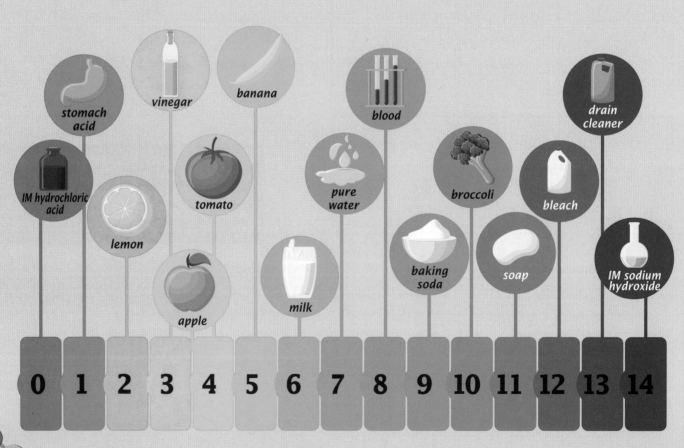

stomach acid

vinegar

banana

blood

drain cleaner

IM hydrochloric acid

tomato

pure water

broccoli

bleach

lemon

apple

milk

baking soda

soap

IM sodium hydroxide

0 1 2 3 4 5 6 7 8 9 10 11 12 13 14

With a little algebra you get ...

$$[OH^-] = \frac{(1.0 \times 10^{-14})}{[1.0 \times 10^{-5}]} = 1.0 \times 10^{-9}$$

You can also do it this way. If H^+ is 1.0×10^{-5}, the pH is 5. $14 - 5 = 9$. So, the $[OH^-]$ is 1.0×10^{-9}. This is one of the ways that pH comes in handy.

Some examples of pH values for common fluids are ...

Common Fluid	pH
Ammonia household cleaner	pH of 12
Blood	pH between 7 and 8
Distilled water	pH of 7
Milk	pH between 6 and 7
Vinegar	pH of 3
Lemon juice	pH between 2 and 3
Stomach acid	pH of 2

Some people say that whenever they try to grow plants, the plants die. Often, this is because of the pH of the soil. A simple kit for measuring the pH of soil samples can be purchased locally where they sell gardening supplies. After checking the pH of the soil, the kits provide information as to how to correct the pH for what you want to grow. Grasses, lilies, and corn grow best in soil pH around 6, and beans and other green vegetables grow best in soil with pH closer to 8.

LABORATORY 20

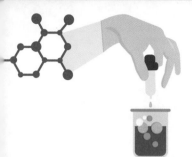

pH AND NaHCO$_3$

REQUIRED MATERIALS

- Vinegar
- pH paper
- Beaker (100 ml)
- Table salt
- Baking soda
- Fruit juice
- Laboratory scoop

PURPOSE

In this exercise, the student is to gain experience in measuring the pH of solutions and to see the variations in pH caused by sodium bicarbonate.

PROCEDURE

1. Take a 40 ml sample of tap water in a 100 ml beaker.

    A. Test the pH of the water using pH paper. (Dip a strip of pH paper into the water and remove it.)

    B. Match the color of the pH paper to the chart on the pH paper container and determine the pH of the tap water.

    C. Record your result.

2. Take a 40 ml sample of vinegar in a 100 ml beaker.

    A. Test the pH of the vinegar and record your result.

    To interpret the results, the pH is the negative value of the logarithm of the [H^+] (acid) concentration. Consider this example. If the pH is 7.0, the [H^+] (molarity of the H^+) is 10^{-7} M. The H in pH stands for the acid (H^+). If the pH turned out to be 5, the [H^+] is 10^{-5} M. 10^{-5} is a larger number than 10^{-7} (0.00001 is larger than 0.0000001), so there is 100 times more acid in the solution with pH 5 than in pH 7. The smaller the pH, the more acid is present.

    What is the pH of your vinegar? What is the [H^+] of the vinegar?

    B. Add 1 gram of NaCl (table salt) to the vinegar and check its pH. Did it change?

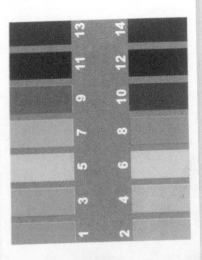

C. Add 1 gram of $NaHCO_3$ (baking soda) to the vinegar and check its pH. Did it change?

The $NaHCO_3$ comes apart in water as ...

$$NaHCO_3 \rightarrow HCO_3^- + H^+$$

The bicarbonate ion (HCO_3^-) can absorb some of the acid becoming carbonic acid as ...

$$HCO_3^- + H^+ \rightarrow H_2CO_3$$

Should this increase or decrease the amount of acid (H^+) in the vinegar?

Compare your measurements of the pH of the vinegar before and after you added the sodium bicarbonate. Did the pH go up or down? Did the amount of acid in the vinegar increase or decrease? Is this what you expected?

3. Take a 40 ml sample of fruit juice in a 100 ml beaker.

A. Test the pH of the fruit juice and record your result, along with the kind of juice you used.

B. Add a gram of $NaHCO_3$ to the juice and check its pH.

What is the new pH? Is this what you expected? What is the H^+ concentration of the juice before you added the $NaHCO_3$? What is the $[H^+]$ of the juice after you added the $NaHCO_3$? What did the $NaHCO_3$ do?

If you spilled some acid, would it be a good idea to clean it up with some baking soda ($NaHCO_3$)?

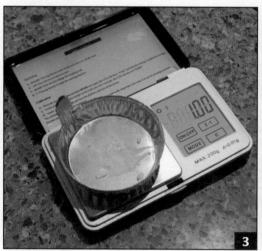

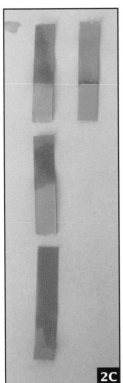
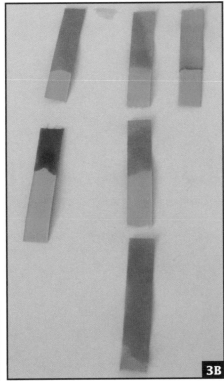

2C

3B

The concept of H$^+$ being an acid and OH$^-$ being a base was developed in the 1880s by the Swedish chemist Svante Arrhenius (1859–1927).

The Danish chemist Johannes Brønsted and the English chemist Thomas Lowry separately in 1923 defined an acid as a proton donor and a base as a proton acceptor. These are called Brønsted-Lowry acids and bases.

The H$^+$ is a proton (no neutron or electron). The acetate ion (CH$_3$COO$^-$) is a proton acceptor, so it is a base.

$$CH_3COO^- + H^+ \longleftrightarrow CH_3COOH \text{ (acetic acid in vinegar)}$$

For water $\quad$ $H^+ + OH^- \longleftrightarrow H_2O$
$\qquad\qquad$ Acid $\quad$ Base

For acetic acid $\quad H^+ + CH_3COO^- \longleftrightarrow CH_3COOH$
$\qquad\qquad$ Acid $\qquad$ Base

The Brønsted-Lowry concept of acids and bases could also apply to gases; whereas, the Arrhenius concept only works for liquid solutions. For example …

$$HCl \text{ (g)} + NH_3 \text{ (g)} \rightarrow NH_4Cl \text{ (solid)}$$

According to Brønsted and Lowry, the acid is HCl and the base is NH$_3$. HCl donates a proton (H$^+$) to NH$_3$, which becomes NH$_4^+$.

Thomas Lowry

CHAPTER 21

ACIDS AND BASES II

OBJECTIVES AND VOCABULARY

The Learning Objectives are for the students to gain an understanding (as evidenced by their performance on the quiz) of:

1. Finding the $[H^+]$ and the $[OH^-]$ from the pH
2. Interpreting the results of an acid-base titration.

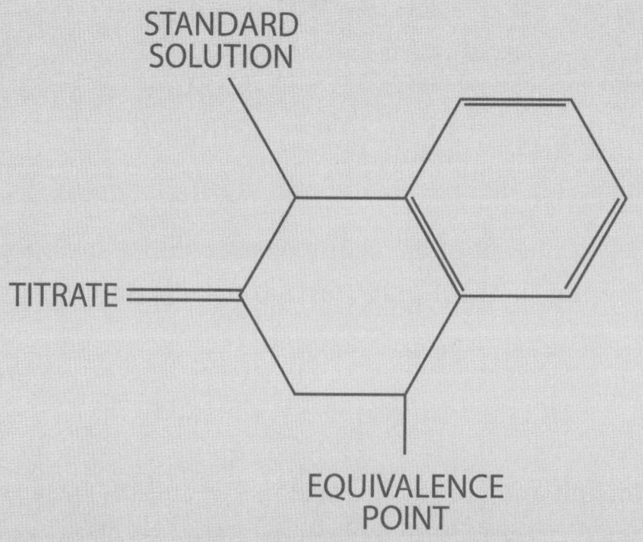
STANDARD
SOLUTION
TITRATE
EQUIVALENCE
POINT

One of the challenges of acid base chemistry is to determine the acid and base content of a solution. When you know the pH of a solution, you can calculate the [H⁺] and the corresponding [OH⁻] of the solution.

As a review, what is the $[H^+]$ of a solution whose pH = 4? The $[H^+]$ for the solution is 1.0×10^{-4} M (moles/liter)? The 10^{-4} comes from the pH of 4.

What is the $[OH^-]$ of the same solution?

$$K_W = [H^+][OH^-] = [1.0 \times 10^{-4}] \, [OH^-] = 1.0 \times 10^{-14}$$

How comfortable you may or may not be with this exercise depends upon your level of understanding of math. Even though this is not a math lesson, it shows how important math can be. If you struggle with the math here, look at the pattern as to how it is done.

$$[OH^-] = \frac{(1.0 \times 10^{-14})}{[1.0 \times 10^{-4}]} = 1.0 \times 10^{-10} \text{ M}$$

Notice that 10^{-14} divided by 10^{-4} is 10^{-10}. You subtract 4 from 14 and get 10.

The exponents of 10 (the logarithms) of the $[H^+]$ and the $[OH^-]$ always add up to 14.

$$\text{If } [H^+] = 1.0 \times 10^{-6} \text{ M, then } [OH^-] = 1.0 \times 10^{-8} \text{ M}$$

$$\text{If } [OH^-] = 1.0 \times 10^{-9} \text{ M, then } [H^+] = 1.0 \times 10^{-5} \text{ M}$$

A little test of your math skills ...

Which is the greater concentration 1.0×10^{-8} M or 1.0×10^{-5} M?

The answer is 1.0×10^{-5} M because ...

$$1.0 \times 10^{-8} = \frac{1}{10^8} = \frac{1}{100,000,000} \text{ and}$$

$$1.0 \times 10^{-5} = \frac{1}{10^5} = \frac{1}{100,000}$$

and 1/100,000 is a much larger number than 1/100,000,000 just like 1/10 is a much larger number than 1/1,000.

If you have a pretty good math background, try this one. If your math is not at this level, do not panic. Do your best to follow the steps. It will become more important later on.

What is the $[H^+]$ of a solution whose pH is 6.7?

$$[H^+] = 10^{-6.7} \text{ M} = 10^{.3} \times 10^{-7}$$

Using a calculator find the number whose log (logarithm) is .3. You will come up with 2. You could also say that the antilog of .3 is 2. You find this on the calculator by entering 3 and then pushing the inv (inverse) key followed by the log key. This means that ...

$$[H^+] = 2.0 \times 10^{-7} \text{ M (the 2 is the same as } 10^{.3})$$

If this is beyond your current level of math, do not worry. It is optional at this point.

Another method used well before calculators and still widely used is to **titrate** the solution in the laboratory. This is sometimes used to test the accuracy of a pH meter. If you have a strong acid like HCl or a strong base like NaOH, the HCl releases all of its H^+ ions and the NaOH releases all of its OH^- ions. A pH meter can detect all of the available H^+ ions, so it gives you an accurate concentration of H^+ ions.

But if you have a weak acid like acetic acid in vinegar or a weak base like ammonia, most of the H$^+$ ions and OH$^-$ ions remain tied up in the acid or base molecules. The pH meter will only detect the H$^+$ ions and OH$^-$ ions that are released and the rest remain undetected. Titrating forces the CH$_3$COOH (acetic acid) to release all of its H$^+$ ions and the NH$_4$OH (ammonium hydroxide) to release all of its OH$^-$ ions. This gives you the total CH$_3$COOH and NH$_4$OH concentrations.

When you titrate a solution of CH$_3$COOH, you gradually add a strong base (OH$^-$) that gradually neutralizes the H$^+$ ions. As the H$^+$ ions are combined with OH$^-$, the CH$_3$COOH releases more H$^+$ ions until it has released all of them. The OH$^-$ neutralizes the H$^+$ ions as

$$H^+ + OH^- \rightarrow H_2O$$

When an acid reacts with a base, water and salt are produced.

$$HCl + NaOH \rightarrow H_2O + NaCl$$

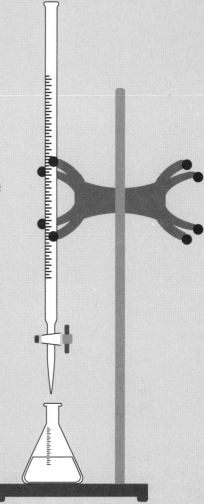

The purpose of this type of titration is to determine the concentration of an acid or base whose concentration is unknown. In this example, you would start with an acid solution whose [H$^+$] is unknown. Add 100 ml (milliliters) of the acid to a flask. Add a few drops of an acid base indicator that changes color when the H$^+$ ions are all combined with OH$^-$ ions. A common indicator is phenolphthalein that turns pink from colorless when the solution goes from acidic to basic. This is a good time to keep things out of your mouth. Phenolphthalein is a strong laxative. Slowly add the NaOH solution drop by drop until the color of the solution turns from colorless to pink. When the critical drop is added, the solution turns pink and stays pink. All of the original H$^+$ ions have combined with the OH$^-$ ions forming H$_2$O. Now you can calculate how many moles of OH$^-$ ions were added, which tells you how many moles of H$^+$ ions were present to begin with.

Titration set up

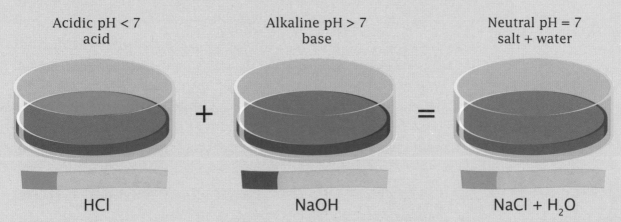

Acidic pH < 7
acid

Alkaline pH > 7
base

Neutral pH = 7
salt + water

HCl

NaOH

NaCl + H$_2$O

According to the equation $H^+ + OH^- \rightarrow H_2O$, there is 1 H^+ ion in the original solution for every OH^- ion added.

Let's say 28.0 ml of 1 M NaOH was added to the flask containing the acid solution. You know the concentration of the NaOH being added but not the concentration of HCl in the flask. The 28.0 ml is 0.028 liter. The moles of NaOH added are …

$$0.028 \text{ l (liter)} \times 1 \text{ mole/liter} = 0.028 \text{ mole of NaOH}$$

There is 1 OH^- ion in each NaOH molecule, so 0.028 mole of OH^- ions is added to the solution. Therefore, there was 0.028 mole of H^+ ions originally in the flask. That is 0.028 mole of H^+ ions in 100ml or 0.1 liter. This gives the original concentration of HCl as …

$$0.028 \text{ mole} / 0.1 \text{ liter} = 0.28 \text{ M HCl}$$

This is a very reliable technique if followed carefully with several duplicate samples to insure that the results agree with each other.

LAB INSIGHTS

TITRATING

When titrating HCl with NaOH …

$$HCl \rightarrow H^+ + Cl^-$$

$$NaOH \rightarrow Na^+ + OH^-$$

$$H^+ + OH^- \rightarrow H_2O$$

$$Na^+ + Cl^- \rightarrow NaCl$$

$$HCl + NaOH \rightarrow NaCl + H_2O$$

A titration begins with a solution of known concentration (called a **standard solution**) where a reacting solution is added. When all of the reactants in the standard solution have reacted with the reactants in the added solution, the **equivalence point** is reached. This is recognized by a change in an indicator, such as the red cabbage solution.

This can also be done by pipetting the standard solution (such as 1 M HCl) into a flask and slowly adding a NaOH solution of unknown concentration. A magnetic stirring rod can keep the solution mixed and a pH meter can monitor the pH. As an acid, the HCl solution will have a pH less than 7. The pH will gradually rise as the NaOH solution is slowly added. When the pH reaches 7, the equivalence point is reached when all of the HCl is neutralized.

$$HCl + NaOH \rightarrow NaCl + H_2O$$

If you add 25 ml of 1 M HCl to the flask at the beginning, you have ...

$$\frac{1 \text{ mole}}{\text{liter}} \times (0.025 \text{ liter}) = 0.025 \text{ mole of HCl}$$

1 mole of NaOH neutralizes 1 mole of HCl, so if you add 30 ml (0.03 liter) of NaOH to get to the equivalence point, you used 0.025 mole of NaOH (the same as the moles of HCl).

$$\frac{(0.025 \text{ mole})}{(0.03 \text{ liter})} = \frac{0.83 \text{ moles}}{\text{liter}} = 0.83 \text{ M NaOH}$$

Now you know the concentration of the NaOH solution.

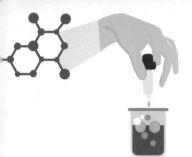

pH OF VARIOUS LIQUIDS

REQUIRED MATERIALS

- pH paper
- Beaker (100 ml) or test tubes
- Assorted household items (See suggested list in Step 2)

INTRODUCTION

Exercise 20 dealt mainly with an acid. In this exercise, you will also deal with some bases.

Water molecules come apart to form an acid and a base.

$$H_2O \rightarrow H^+ \text{ (acid)} + OH^- \text{ (base)}$$

About 1 out of every 10 million (10^7) water molecules comes apart. For every water molecule that comes apart, H^+ and OH^- come together to reform another water molecule. This way they stay in balance and the acid and base do not build up. One out of 10^7 gives $[H^+]$ and $[OH^-]$ concentrations of 10^{-7} M for the acid and the base. This is why a pH of 7 indicates that the solution is neutral or that there is as much acid as base present. When the pH goes below 7, there is more acid than base. For example, at pH 5, the $[H^+]$ is 10^{-5} M. As we saw in exercise 20, 10^{-5} is a larger number than 10^{-7}. The difference is 100 times. So, the $[H^+]$ in 10^{-5} M is 100 times more concentrated than when $[H^+]$ is 10^{-7} M. Likewise, a solution of pH 3 has 10^4 or 10,000 times more acid than a solution of pH 7.

PURPOSE

This exercise is to provide practice in testing the pH of various liquids and for the student to be aware of the pH of various liquids encountered in everyday life.

PROCEDURE

1. Before beginning: prepare a chart like the one below to report your results (see Teacher Guide):

Liquid	pH	[H$^+$]	pOH	[OH$^-$]	Acid / Base

2. Take a sample of each of the following solutions in a test tube or 100 ml beaker, and check the pH with pH indicator paper. If you do not have something on this list, you can substitute something similar. Record your results.

   A. Dish soap in water

   B. Coffee or tea

   C. Milk

   D. Soda

   E. Ketchup

   F. Egg white dissolved in water

   G. Salad dressing

   H. Window cleaner

   I. An antacid ground up and dissolved in water

   J. Hand soap in water

   K. Mouthwash

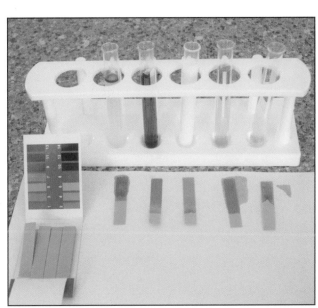

3. For each sample, indicate the [H^+].

When a water molecule comes apart, it forms a base, as well as an acid.

$$H_2O \rightarrow H^+ + OH^-$$

In a neutral solution, [H^+] = [OH^-]. And when they are balanced, [H^+] [OH^-] = 10^{-14}. To find the concentration of base from the pH, use 14 – pH = pOH. If the pH is 5, the pOH is (14 – 5) or 9 and the [OH^-] is 10^{-9}. The liquids with a pH greater than 7 are basic (meaning that they have more [OH^-] than [H^+]).

4. Your chart should show the identity of the liquid, the pH, and the [H^+]. Add to the chart the pOH, which is 14 – pH. If the pH is 8, the pOH is 14 – 8 = 6. If the pH is 4, the pOH is 14 – 4 = 10.

5. From the pOH, figure the [OH^-]. For example, if pOH is 9, the [OH^-] = 10^{-9} M. If the pOH is 3, the [OH^-] is 10^{-3} M. Add the [OH^-] values to your chart.

6. Finally, for each liquid, indicate if it is acidic (pH below 7) or basic (pH above 7).

CONJUGATE ACIDS AND BASES

LAB INSIGHTS

The word "conjugate" means joined together as a pair. For the reaction ...

$$HCl \quad + \quad H_2O \quad \rightarrow \quad Cl^- \quad + \quad H_3O^+$$

Acid   Base   Conjugate base   Conjugate acid

The conjugate base is formed by removing a proton from an acid. The conjugate acid is formed by adding a proton to the base.

BASES ACQUIRE PROTONS

Bases acquire protons. Here is a scale of increasing strengths of bases.

$$Cl^- \rightarrow HSO_4^- \rightarrow SO_4^{--} \rightarrow HCO_3^- \rightarrow CO_3^{--} \rightarrow OH^- \rightarrow H^-$$

Weak Bases $\rightarrow$ Stronger Bases

LAB INSIGHTS

Consider this reaction ...

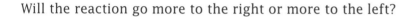

$$HSO_4^- + CO_3^{--} \leftrightarrow SO_4^{--} + HCO_3^-$$

Acid   Base   Conjugate base   Conjugate acid

Will the reaction go more to the right or more to the left?

CO_3^{--} is a stronger base than SO_4^{--} and is more likely to get the protons than SO_4^{--}, so the reaction is favored to go to the right.

Consider these acids listed from the weakest to the strongest:

$$HPO_4^- \rightarrow HCO_3^- \rightarrow H_2PO_4^- \rightarrow H_2CO_3 \rightarrow HSO_4^-$$

Weak acids $\rightarrow$ Strong acids

In the above reaction, the acids are HSO_4^- and HCO_3^-.

HCO_3^- is weaker than HSO_4^-, so the HCO_3^- will come apart easier, causing the reaction to go more to the right than to the left. This is the same conclusion as when considering the weak and strong bases.

It is important that all of the chemical reactions that occur anywhere in the universe, including those within our cells, proceed in the correct direction. When considering all of the molecules and reactions that must occur all the time, this was all spoken into being by our great and gracious God. Why would we ever question His wisdom and direction?

CHAPTER 22

WEAK ACIDS AND BASES

OBJECTIVES AND VOCABULARY

The Learning Objectives are for the students to gain an understanding (as evidenced by their performance on the quiz) of:

1. Writing out dissociation constants for weak acids and weak bases

2. Writing out solubility products for compounds of limited solubilities

3. Showing how solutions reestablish their equilibrium.

TRIPROTIC ACID

SOLUBILITY PRODUCT

MONOPROTIC ACID

DIPROTIC ACID

R emember that the strong acids release all of their H⁺ ions and strong bases release all of their OH⁻ ions.

$$HCl \rightarrow H^+ + Cl^-$$

$$NaOH \rightarrow Na^+ + OH^-$$

Weak acids and bases, however, release only some of the H⁺ ions and OH⁻ ions. The equations for the breakdown of acetic acid and ammonium hydroxide are reversible meaning that they reach equilibrium when the products recombine to reform reactants just as fast as the products are formed. Even though the reaction is going back and forth, the concentrations of reactants and products are not changing. It is similar to spending money out of your checking account at the same rate that you are putting money back into the account. This way the total amount of money in the account stays the same.

For acetic acid

$$CH_3COOH \leftrightarrow CH_3COO^- + H^+$$

For ammonium hydroxide

$$NH_4OH \longleftrightarrow NH_4^+ + OH^-$$

When dealing with a strong acid like HCl, you can tell how much H^+ you have by how much HCl you have. But you cannot do this with a weak acid. The amount of H^+ released from CH_3COOH is much less than the amount of CH_3COOH present. To find out how much H^+ is released in a solution of CH_3COOH, you need to use the dissociation constant of acetic acid. Remember that the dissociation constant of water is ...

$$H_2O \longleftrightarrow H^+ + OH^-$$

$$K_W = [H^+][OH^-] = 1.0 \times 10^{-14}$$

The dissociation constant of water only involves the products H^+ and OH^-. The amount of H_2O is not part of the equation because water is the solvent. In other words, if the amount of water changes, the concentrations of H^+ and OH^- will adjust to match it. But it is different for acetic acid. Acetic acid is not the solvent; water is.

The dissociation constant of acetic acid is written as ...

$$CH_3COOH \longleftrightarrow CH_3COO^- + H^+$$

$$K_a = [CH_3COO^-][H^+] / [CH_3COOH] = 1.76 \times 10^{-5}$$

The "$_W$" in K_W stands for water and the "$_a$" in K_a stands for acid. For the weak base ammonium hydroxide, K_b is used for the dissociation constant where "$_b$" stands for base.

$$NH_4OH \longleftrightarrow NH_4^+ + OH^-$$

$$K_b = [NH_4^+][OH^-] / [NH_4OH]$$

Notice that the expressions for the dissociation constants always have the concentrations of the products times each other divided by the concentrations of the reactants. If there were more than one reactant, they would be multiplied by each other.

An important skill to develop in life is to be able to remember what you have learned and be able to use it later. In the sciences, especially, you have to go back and review what you learned before and use it as you go along. This is a skill that you need to take seriously and work hard at it. This skill is critical to your growth and will help you in anything you do. This is also important for the study of the Bible because different passages of Scripture have to come together for understanding.

Look at this equation for sulfuric acid (found in car batteries). How is it different from those that you have already studied?

$$H_2SO_4 \longleftrightarrow 2H^+ + SO_4^{--}$$

The H_2SO_4 releases 2 H^+ ions while those that you have looked at so far only release 1 H^+. The equilibrium constant (same as dissociation constant) for H_2SO_4 is ...

$$K_a = [H^+]^2[SO_4] / [H_2SO_4]$$

Notice that the 2 H^+ is expressed as $[H^+]^2$. There are 2 H^+ product molecules whose concentrations are multiplied by each other becoming $[H^+]^2$.

An acid that releases 1 H^+ ion for each acid molecule (like HCl) is called a **monoprotic acid**. An acid, like H_2SO_4, that releases 2 H^+ ions from each acid molecule is called a **diprotic acid**. Phosphoric acid, H_3PO_4, is a **triprotic acid**.

$$H_3PO_4 \longleftrightarrow 3\,H^+ + PO_4^{-3} \text{ (This is PO}_4 \text{ with a } -3 \text{ charge)}$$

$$K_a = [H^+]^3[PO_4^{-3}] / [H_3PO_4]$$

Diprotic and triprotic acids break down in steps. This can be seen in the example of carbonic acid (H_2CO_3).

$$H_2CO_3 \longleftrightarrow H^+ + HCO_3^- \text{ (bicarbonate ion)}$$

$$HCO_3^- \longleftrightarrow H^+ + CO_3^{--} \text{ (carbonate ion)}$$

Together they look like ...

$$H_2CO_3 \longleftrightarrow 2\,H^+ + CO_3^{--}$$

$$K_a = [H^+]^2[CO_3^{--}] / [H_2CO_3]$$

This is produced when CO_2 in the atmosphere combines with water.

$$CO_2 + H_2O \longleftrightarrow H_2CO_3$$

A **monoprotic base** means that a molecule gives up 1 OH^- such as NaOH $\rightarrow Na^+ + OH^-$. That is where the "mono" part of the word comes from. A **diprotic base** gives up 2 OH^- such as $Ca(OH)_2 \rightarrow Ca^{++} + 2\ OH^-$. Look at the example of calcium phosphate $Ca_3(PO_4)_2$.

$$Ca_3(PO_4)_2 \longleftrightarrow 3\ Ca^{++} + 2\ PO_4^{-3}$$

$$K_{sp} = [Ca^{++}]^3[PO_4^{-3}]^2$$

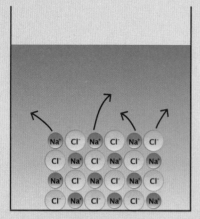

Salt is initially put into the water and begins dissolving.

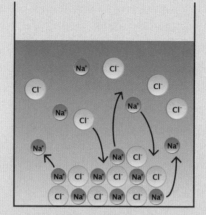

Salt continues to dissolve; however, dissolved ions will also precipitate. Because the salt dissolves faster than its ions precipitate, the net movement is towards dissolution.

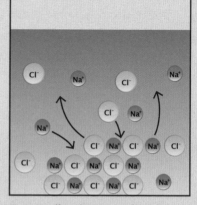

Eventually, the rate of dissolution will equal the rate of precipitation. The solution will be in equilibrium, but the ions will continue to dissolve and precipitate.

In the case of $Ca_3(PO_4)_2$, the equilibrium constant is called a **solubility product** because instead of dealing with an acid or a base, it is dealing with the solubility of a salt that does not readily dissolve in water. Notice that the salt $Ca_3(PO_4)_2$ is not part of the solubility product because it settles (precipitates) to the bottom of the container and is not part of the solution.

A very interesting prediction can be made from equilibrium constants. If you add something to a solution that changes some of the concentrations, changing the ratio of products and reactants, the reaction will proceed forwards or backwards to change the concentrations so that the ratio of products and reactants goes back to its original value. This may be a bit confusing. The example of carbonic acid will help.

$$H_2CO_3 \leftrightarrow 2\ H^+ + CO_3^{--}$$

When CO_2 dissolves in water, as happens in the oceans, lakes, and rivers, carbonic acid is formed.

$$H_2O + CO_2 \leftrightarrow H_2CO_3$$

The equilibrium of carbonic acid is disrupted because $[H_2CO_3]$ increases in value. Some of the H_2CO_3 breaks down increasing the $[CO_3^{--}]$ and $[H^+]$ so that the ratio of the products divided by the reactants comes back to its original value.

$$K_a = [H^+]^2[CO_3^{--}]\ /\ [H_2CO_3]$$

When the H_2CO_3 is increased, the fraction is divided by a larger number, reducing the overall value of the fraction. Then some of the H_2CO_3 breaks apart (decreasing the $[H_2CO_3]$) increasing the values of $[H^+]$ and $[CO_3^{--}]$, which restores the fraction to its original value of K_a. This is why it is called a constant. If it is changed, it will come back to its original value.

Consider ammonium hydroxide ...

$$NH_4OH \longleftrightarrow NH_4^+ + OH^-$$

If something reacts with the NH_4^+ reducing its concentration, what happens?

$$K_b = [NH_4^+][OH^-] / [NH_4OH]$$

If $[NH_4^+]$ decreases, the fraction becomes smaller just like 1/3 is smaller than 2/3. Some of the NH_4OH then breaks apart so that the value on the bottom of the fraction decreases. The $[NH_4^+]$ and $[OH^-]$ are increased, raising the top of the fraction until the overall fraction comes back to its original value.

This is a very handy principle for predicting the extent and direction of many chemical reactions.

In this study, notice the patterns of what happens. This way you could handle many situations that you have never seen before. That is far more valuable than trying to memorize many different reactions.

At Mammoth Hot Springs in northwest Wyoming in Yellowstone National Park, many solutes (Mg^{++}, Ca^{++}, Fe^{++}, CO_3^- and SO_4^{--}) are dissolved in the underground water because of its acidity and high temperature. H_2CO_3 (carbonic acid) helps keep the solutes dissolved. When the water flows out onto the surface, carbonic acid breaks down into carbon dioxide and evaporates, reducing the acidity and increasing the availability of electrons for many of the solutes.

$$2 H^+ + CO_3^{--} \rightarrow H_2CO_3 \rightarrow H_2O + CO_2$$

This causes many mineral solutes to precipitate (come out of solution) and add to the travertine terraces.

When minerals are neutral in charge (*i.e.* Mg^0, Ca^0 and Fe^0), they are solid. When they are oxidized, they go into solution (*i.e.* $Mg^0 \rightarrow Mg^{++} + 2\ e^-$).

Mammoth Hot Springs

215

ACID-BASE pH AND TITRATION

REQUIRED MATERIALS

- Erlenmeyer flask
- Beaker (250 ml)
- Beaker (100 ml)
- Burette (50 ml)
- Test tubes
- Stirring rod
- pH paper
- Barium hydroxide
- Vinegar
- Red cabbage (1/2 head)
- Cooking pot
- Strainer
- Distilled water
- Weighing boat
- Scale
- Laboratory scoop

INTRODUCTION

An alternative method of determining whether a solution is acidic or basic or neutral is to use a liquid indicator. This is a liquid that has a different color in an acid versus in a base.

A homemade indicator is the juice from red cabbage. The dark purplish liquid contains molecules called anthocyanins. They are part of a group of molecules called flavonoids. Most pH indicators change color by absorbing H^+ ions from acids. In basic solutions, they give up their H^+ ions to the base. The anthocyanins, however, absorb OH^- ions from basic compounds.

PURPOSE

This exercise demonstrates equilibrium that exists between the acid released from a weak acid and the acid retained by the weak acid.

PROCEDURE

NOTE: Please wear protective equipment while conducting this lab.

1. Prepare a red cabbage juice indicator solution.

   A. Grate or chop up ½ head of red cabbage.

   B. Place the pieces in a pot of water and boil it for 20 to 30 minutes until the liquid turns a dark purplish color.

   C. Strain 50 ml into a 100 ml beaker and set it aside to cool for later use.

2. Take a 50 ml sample of vinegar (about 5 percent acetic acid) and determine its pH using pH paper.

3. Make a 0.1 M solution of barium hydroxide ($Ba(OH)_2$).

   A. Measure 100 ml of distilled water into a 250 ml beaker.

   B. Measure out 1.7 g of $Ba(OH)_2$. (The strong base $Ba(OH)_2$ has a molecular mass of 171 grams/mole.)

   C. Stir the $Ba(OH)_2$ into the distilled water until it is completely dissolved.

   D. Take a 5 ml sample of the 0.1 M $Ba(OH)_2$ solution and add a few drops of your red cabbage indicator.

   What is the color of your indicator in this basic solution?

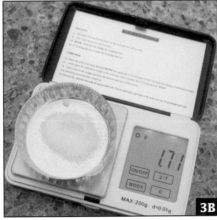

3A 3B 3C

E. Add a few drops of your indicator to 5 ml of vinegar.

What is the color of your indicator in an acidic solution?

F. Add a few drops of your indicator to 5 ml of distilled water.

What is the color of your indicator when a few drops are placed in 5 ml of distilled water?

4. Add 10 ml of vinegar to an Erlenmeyer flask.

5. Add a few drops of your red cabbage indicator to the flask.

The acetic acid in the vinegar releases some of its H^+ ions. The reaction whereby acetic acid releases the H^+ ions is …

$$H_3CCOOH \longleftrightarrow H_3CCOO^- + H^+$$

There is an equilibrium where the ratio

$[H_3CCOO^-][H^+] / [H_3CCOOH]$ is a constant.

If some of the H^+ ions are removed, the ratio is decreased. To restore it to its original value, more of the H_3CCOOH breaks down releasing more H_3CCOO^- and H^+. If more of the H^+ is removed, the H_3CCOOH will continue to break apart until there is no more left.

The process of gradually removing the H^+ ions until they are gone is called titration. Most of the time, 0.1 M NaOH is gradually added to a weak acid until the solution stays basic (there is no more acid left). NaOH is very caustic, and it is very hydroscopic (absorbs water), making it difficult to weigh. In this exercise, use the 0.1 M $Ba(OH)_2$ solution instead. $Ba(OH)_2$ is a strong base, meaning that it will release all of its OH^- ions. There are 2 OH^- ions in every $Ba(OH)_2$ molecule, so 0.1 M $Ba(OH)_2$ yields 0.2 M OH^-.

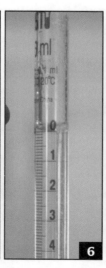

6. Prepare the burette with the $Ba(OH)_2$ solution. (See Burette Preparation under Laboratory Procedures on page 275])

7. Gradually add the $Ba(OH)_2$ solution to the Erlenmeyer flask with 50 ml of vinegar from Steps 3–4 until the red cabbage indicator turns to the color that you expect for a basic solution.

8. Check the pH of the final solution with your pH paper. It should be basic (a pH just above 7).

9. Record the ml of $Ba(OH)_2$ solution added.

10. Convert the milliliters used to liters.

   For example, 43 ml would be 0.043 liter.

11. Calculate the number of moles of OH^- you added to the flask.

   Multiply the volume by the molarity to get the number of moles of OH^- added to neutralize the H^+ ions (0.043 liter x 0.2 mole/liter = 0.009 mole).

$$0.043 \text{ liter } \times 0.2 \frac{\text{mole}}{\text{liter}} = 0.009 \text{ mole}$$

   This equals the amount of H^+ ions that were neutralized.

12. Divide the number of moles of H^+ by the volume (50 ml or 0.05 liter), and you get the moles per liter of the acetic acid in the vinegar.

   The pH that you measured with the pH paper can be converted to molarity as in this example — if the pH = 6.0 the $[H^+] = 10^{-6}$. How does this molarity compare to what you get from the titration? The titration should give a higher molarity of H^+ ions because it pulls all the H^+ off the acetic acid molecules and the pH paper does not.

CHAPTER 23

BUFFERS

OBJECTIVES AND VOCABULARY

The Learning Objectives are for the students to gain an understanding (as evidenced by their performance on the quiz) of:

1. How pH buffers keep the pH constant

2. Buffering capacity

3. How to prepare a buffer for a particular pH

4. Bicarbonate-carbonate buffer

5. The phosphate buffer.

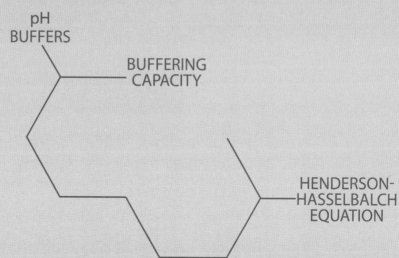

pH
BUFFERS

BUFFERING
CAPACITY

HENDERSON-
HASSELBALCH
EQUATION

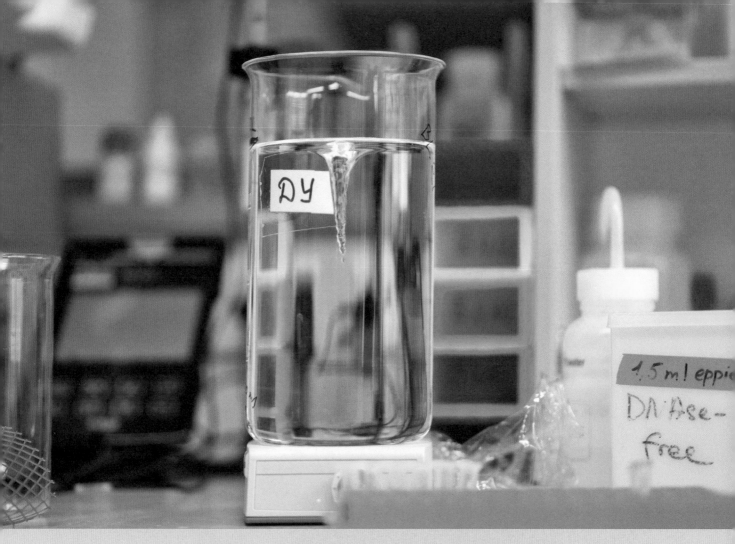

Chapter 20 introduced the concept of chemical equilibrium. An important application lies in the need to maintain constant pH. The pH of our blood has to stay between 7.35 and 7.45. If the pH goes below 7 or above 7.8, death occurs. A change in pH destroys proteins that form structures within cells and the many enzymes that control all of the chemical reactions in cells. This is a very narrow range of pH. It could make us feel rather nervous and fragile. But God through His grace created molecules to control our pH, called pH buffers. You have already studied buffers in previous lessons. They were not called buffers.

The word buffer means something that prevents extreme changes. The bicarbonate – carbonate buffer was studied in Chapter 22. In a chemistry lab, sodium bicarbonate ($NaHCO_3$) is kept handy in case there is either an acid or a base spill. It is very effective at neutralizing both.

$$NaHCO_3 \rightarrow Na^+ + HCO_3^-$$
$$HCO_3^- + H^+ \text{ (excess acid)} \rightarrow H_2CO_3$$
$$HCO_3^- + OH^- \text{ (excess base)} \rightarrow CO_3^{--} + H_2O$$

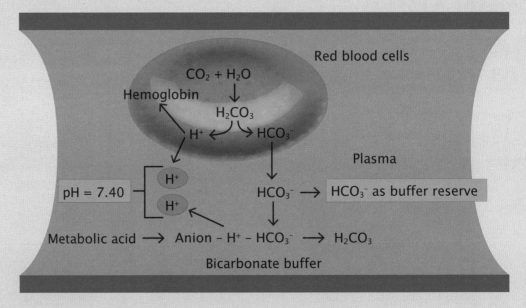

H_2CO_3 (carbonic acid) is formed when H_2O combines with CO_2.

$$CO_2 + H_2O \longleftrightarrow H_2CO_3 \qquad \text{carbonic acid}$$
$$H_2CO_3 \longleftrightarrow H^+ + HCO_3^- \quad \text{bicarbonate ion}$$
$$HCO_3^- \longleftrightarrow H^+ + CO_3^{--} \quad \text{carbonate ion}$$

If $[H^+]$ increases lowering the pH, the excess H^+ ions can be absorbed by HCO_3^- to form H_2CO_3 and the CO_3^{--} ion to form HCO_3^-.

$$H^+ + HCO_3^- \longleftrightarrow H_2CO_3$$
$$H^+ + CO_3^{--} \longleftrightarrow HCO_3^-$$

If $[OH^-]$ increases raising the pH, excess OH^- ions can be absorbed by H^+ released from H_2CO_3 or HCO_3^-.

$$H_2CO_3 \longleftrightarrow H^+ + HCO_3^- \quad \text{and} \quad H^+ + OH^- \longleftrightarrow H_2O$$
$$HCO_3^- \longleftrightarrow H^+ + CO_3^{--} \quad \text{and} \quad H^+ + OH^- \longleftrightarrow H_2O$$

The amount of H^+ and OH^- that can be absorbed depends upon the amount of H_2CO_3 and HCO_3^- available. This is the **buffering capacity** of the solution. As long as we metabolize food, producing CO_2, we have a supply of these important buffers. It is interesting that what is sometimes seen as harmful is really another example of God's boundless grace.

Another important buffering system in our bodies is the **phosphate buffering system**. This involves orthophosphoric acid H_3PO_4. This is a triprotic weak acid.

$$H_3PO_4 \leftrightarrow H^+ + H_2PO_4^-$$
$$H_2PO_4^- \leftrightarrow H^+ + HPO_4^{--}$$
$$HPO_4^{--} \leftrightarrow H^+ + PO_4^{-3}$$

Excess H^+ can be absorbed by $H_2PO_4^-$, HPO_4^{--} and PO_4^{-3} .

$$H^+ + H_2PO_4^- \leftrightarrow H_3PO_4$$
$$H^+ + HPO_4^{--} \leftrightarrow H_2PO_4^-$$
$$H^+ + PO_4^{-3} \leftrightarrow HPO_4^{--}$$

Excess OH^- can be absorbed by H_3PO_4, $H_2PO_4^-$ and HPO_4^{--}.

$$H_3PO_4 \leftrightarrow H^+ + H_2PO_4^- \quad \text{and} \quad OH^- + H^+ \leftrightarrow H_2O$$
$$H_2PO_4^- \leftrightarrow H^+ + HPO_4^{--} \quad \text{and} \quad OH^- + H^+ \leftrightarrow H_2O$$
$$HPO_4^{--} \leftrightarrow H^+ + PO_4^{-3} \quad \text{and} \quad OH^- + H^+ \leftrightarrow H_2O$$

Ca^{++} and the phosphate ion PO_4^{-3} together form $Ca_3(PO_4)_2$, which is called calcium phosphate in chemistry and apatite in physiology. When the Ca^{++} and/or PO_4^{-3} concentrations in the body fluids fall below a critical level, more Ca^{++} and PO_4^{-3} is released from bone tissue to raise their levels and maintain the phosphate buffering system.

In the laboratory, materials such as proteins and nucleic acids (DNA) require very strict pH control. The bicarbonate buffering system is often used. You cannot just dump in some $NaHCO_3$ (sodium bicarbonate) and assume that it will keep the pH where you want it to be. You first have to decide upon the desired pH and buffering capacity.

To find the concentrations of buffering compounds needed to maintain a desired pH, use the **Henderson–Hasselbalch equation**. To use this equation, the pK is used. This is similar to the pH. It is the negative logarithm of the equilibrium constant (K_a) just like the pH is the negative logarithm of the [H^+]. If the bicarbonate buffering system is used,

$$H_2CO_3 \leftrightarrow H^+ + HCO_3^-$$
$$K_a = [H^+][HCO_3^-] / [H_2CO_3] = 4.47 \times 10^{-7}$$

This can be rewritten as …

$$4.47 = 10^{.65} \text{ because } 0.65 \text{ is the logarithm of } 4.47$$
$$10^{.65} \times 10^{-7} = 10^{-6.35}$$

You can find the logarithm of a number by entering the number into your calculator and pushing the "log" button. On some you push the "log" button first and then enter the number.

With $K_a = 10^{-6.35}$, the pK is 6.35. The logarithm (power of 10) of the K_a is −6.35 and the negative of −6.35 is 6.35.

The Henderson-Hasselbalch equation is ...

$$pH = pK + \log [HCO_3^-] / [H_2CO_3]$$

To maintain a solution at pH = 7.4, the equation becomes ...

$$7.4 = 6.35 + \log [HCO_3^-] / [H_2CO_3]$$

If you subtract 6.35 from both sides of the equation, you get ...

$$7.4 - 6.35 = 1.05 = \log [HCO_3^-] / [H_2CO_3]$$

The next step is to find the antilogarithm of both sides of the equation.

With a calculator, you can find the number whose logarithm is 1.05, which is about 11. This is found by pushing "inv" then "log" and then 1.05 on your calculator. This means that the logarithm of 11 is 1.05 or that $10^{1.05} = 11$.

$$\text{The ratio of } [HCO_3^-] / [H_2CO_3] = 11$$

This means that no matter how much HCO_3_- (bicarbonate) and H_2CO_3 is present, the amount of HCO_3^- divided by the amount of H_2CO_3 has to be 11. The HCO_3^- is usually used in the form of $NaHCO_3$ (sodium carbonate) that comes apart to release the HCO_3^- into the solution. This is a ratio of a salt ($NaHCO_3$) divided by a weak acid (H_2CO_3). The concentration of $NaHCO_3$ has to be 11 times that of H_2CO_3. If you use more $NaHCO_3$, you have to use more HCO_3^-. Using more $NaHCO_3$ and H_2CO_3 increases the buffering capacity. You have to be careful because too much $NaHCO_3$ and H_2CO_3 could have adverse effects. Greater buffering capacity means that more H^+ and OH^- can be controlled.

Do not get discouraged if the math has gotten away from you this time. How well you follow the use of the Henderson-Hasselbalch equation

Bicarbonate

depends upon your math background up to this time, not how smart you are. If need be, your math can catch up later and you can then come back and take another look at this discussion. In the meantime, be sure that you get the basic lesson that buffers absorb the excess acid and base. You do not have to calculate the ratios of salts and weak acids for the level of this study. God does a much better job with math. He created math. He knows exactly how much $NaHCO_3$ and H_2CO_3 to use in your body fluids to hold your pH steady and give enough but not too much buffering capacity. Your study should not discourage you but instead cause you to rejoice in your loving Creator and want to understand Him and His creation even more.

Blood is a good example of a buffered solution. The major buffering system designed by God to keep the pH from varying is the carbonic acid — bicarbonate system.

$$H^+ + HCO_3^- \leftrightarrow H_2CO_3 \leftrightarrow H_2O + CO_2$$

$$\text{Bicarbonate} \qquad \text{Carbonic acid}$$

When our cells produce CO_2, they continually supply pH buffers. If our body fluids become more acidic, excess H^+ is absorbed by HCO_3^-, forming H_2CO_3. If our body fluids become more basic, H_2CO_3 releases more H^+ to neutralize the excess OH^- ($H^+ + OH^- \rightarrow H_2O$).

Oxygen is carried in red blood cells by the protein hemoglobin (Hb). Hb reversibly binds both H^+ and O_2.

$$HbH^+ + O_2 \leftrightarrow HbO_2 + H^+$$

When excess H^+ is present, it reacts with HbO_2, forcing the reaction to the left releasing O_2. You want the O_2 released in the cells but not in the lungs and blood flowing to the cells.

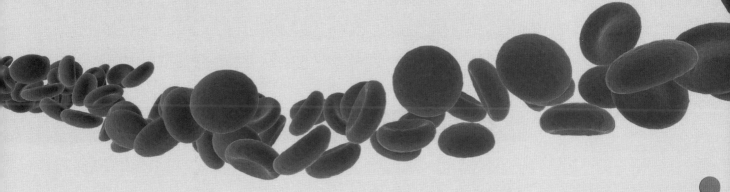

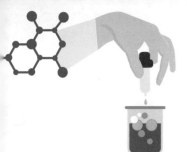

PREPARING AND TESTING BUFFERS

REQUIRED MATERIALS

- Beakers (3x 100 ml)
- Vinegar
- Sodium acetate
- pH paper
- Baking soda on-hand in case of spills
- Toilet bowl cleaner
- Stirring rod
- Weighing boats
- Scale
- Laboratory scoop
- Graduated cylinder (10 ml)
- Eyedropper

INTRODUCTION

There are several buffer systems in our bodies. They are composed of a weak acid and the base of a weak acid. For example, a H_2O molecule comes apart to give H^+ (acid) and OH^- (base). Several buffers are described in the lesson dealing with buffers. In this exercise, the acetate buffering system is used. Acetic acid (in vinegar) comes apart to yield H^+ and the acetate ($H_3C_2O_2^-$) ion. The H^+ is the acid and $H_3C_2O_2^-$ is the base, just as OH^- is the base released from H_2O. The ratio of the base/weak acid determines the pH of the buffer. If extra H^+ is added, the base ($H_3C_2O_2^-$) combines with it, forming more acetic acid ($H_3C_2O_2H$). If extra base (OH^-) is added, the H^+ combines with it to form H_2O. This way, extra acid or base can be absorbed without changing the pH.

PURPOSE

This exercise gives the student experience in preparing a pH buffer and testing its effectiveness.

PROCEDURE

NOTE: Please wear protective equipment while conducting this lab.

1. Pour out 3 samples of 50 ml (milliliters) of vinegar in 100 ml beakers and number them from 1 to 3.

2. Add sodium acetate ($H_3C_2O_2Na$) to the samples and stir until dissolved:

   A. Add 3.4 g (grams) to #1.

   B. Add 6.8 g to #2.

   C. Add 10.2 g to #3.

   The $H_3C_2O_2Na$ is a salt like NaCl. It comes apart in water to yield the acetate ion ($H_3C_2O_2^-$) and the sodium (Na^+) ion.

3. Check the pH of each solution with the pH paper and record the values.

4. To each sample, add ½ an eye dropper of toilet bowl cleaner at a time until the pH becomes lower (more acidic). Test the pH after each ½ dropper of toilet bowl cleaner with pH paper. It is sufficient to estimate ½ of a dropper of toilet bowl cleaner because you do not need to measure an exact amount to know which solution takes more than the others.

   Most toilet bowl cleaner has 9.5 percent hydrogen chloride (hydrochloric acid). This acid is corrosive, so be careful when handling it and protect all surfaces. When the pH changes it should go lower, becoming more acidic.

  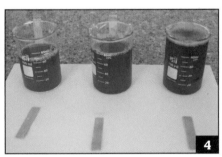

5. Record the number of 1/2-full eyedroppers you need to change the pH.

   If you spill any, pour sodium bicarbonate (baking soda) on it to neutralize it. Then clean it up with water.

   Which sample took the least amount of acid to change the pH? Which took the most? Which of the 3 samples gave you the best buffer?

   One of these combinations is better than the others. It is based upon the Henderson-Hasselbalch equation. The key to a good buffer is the ratio of the concentration of a base divided by the concentration of the weak acid.

CHAPTER 24

CHEMISTRY OF CARBON

OBJECTIVES AND VOCABULARY

The Learning Objectives are for the students to gain an understanding (as evidenced by their performance on the quiz) of:

1. The nature of single carbon bonds

2. How carbon atoms can form millions of different molecules

3. The nature of double carbon bonds

4. The nature of triple carbon bonds.

HYBRIDIZING
ORBITALS

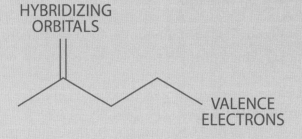

VALENCE
ELECTRONS

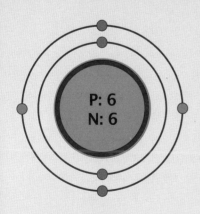

Adam being formed from the dust of the earth.

Carbon is a very special element. It is very rare in the crust of the Earth (0.027 percent). This means that when God created Adam from the dust of the earth, He either gathered carbon atoms from a very wide area or miraculously turned the soil into carbon. Most of the crust of the Earth is composed of silicon dioxide (SiO_2), which is quartz. Ground-up quartz is sand. If life evolved from what was available, we should be made of quartz. Awkward! Carbon as CO_2 is only found in the atmosphere at 375 ppm (parts per million). That is 0.0375 percent. That is also rare.

If life evolved from what was available, we should be made of quartz.

Carbon is the building block of all biological molecules. Proteins, lipids (fats), nucleic acids (DNA and RNA), and carbohydrates are mostly composed of carbon with some oxygen, nitrogen, phosphorus, and sulfur. But herein lies a mystery. A carbon atom has 6 protons and 6 electrons. The electron configuration of a carbon atom is $1s^2\ 2s^2\ 2p_x^1\ 2p_y^1$. This means that carbon can only form 2 covalent bonds (with its 2 unpaired electrons). The only compound that does that is CO (carbon monoxide), which is a deadly poison. But carbon forms well over a million compounds in our bodies.

P: 6
N: 6

Carbon
$1s^2\ 2s^2\ 2p_x^1\ 2p_y^1$

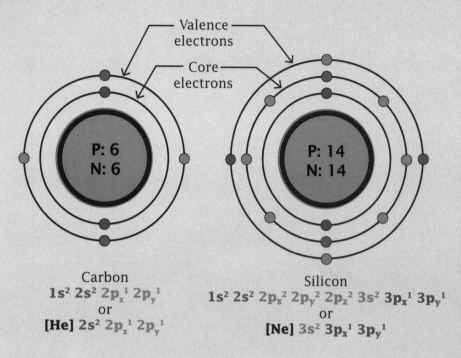

Carbon
$1s^2\ 2s^2\ 2p_x^{\ 1}\ 2p_y^{\ 1}$
or
[He] $2s^2\ 2p_x^{\ 1}\ 2p_y^{\ 1}$

Silicon
$1s^2\ 2s^2\ 2p_x^{\ 2}\ 2p_y^{\ 2}\ 2p_z^{\ 2}\ 3s^2\ 3p_x^{\ 1}\ 3p_y^{\ 1}$
or
[Ne] $3s^2\ 3p_x^{\ 1}\ 3p_y^{\ 1}$

Look at the periodic table. He (helium) has the electron configuration $1s^2$, which makes it for the most part unreactive as a Noble Gas. The electron configuration of carbon can also be written as [He] $2s^2\ 2p_x^{\ 1}\ 2p_y^{\ 1}$. The $1s^2$ electrons (shown by [He]) are the core electrons. The $2s^2\ 2p_x^{\ 1}\ 2p_y^{\ 1}$ electrons are the **valence electrons** because they have the highest principle quantum number. As another example, consider silicon. The electron configuration of Si with 14 electrons is $1s^2\ 2s^2\ 2p^6\ 3s^2\ 3p_x^{\ 1}\ 3p_y^{\ 1}$. This can be rewritten as [Ne] $3s^2\ 3p_x^{\ 1}\ 3p_y^{\ 1}$. Do you see the resemblance to carbon? Remember that the elements in the same group (column) of the periodic table are similar to each other. This is why Si was mentioned earlier.

We still have a problem with C (carbon). How does a carbon atom form more than 2 covalent bonds? God designed a rather clever plan to handle this problem called **hybridizing orbitals**. A hybrid is a combination of 2 or more different things — like a car that has a gasoline motor and an electric motor. In the molecule methane (CH_4), also called natural gas (that you use to cook, heat water and even run some vehicles), carbon forms 4 equal covalent bonds with 4 hydrogen atoms. Think of the shape of a pyramid (like we did with the water molecule) with the C atom in the middle and an H atom at each of the corners.

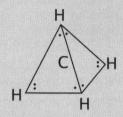

How do we get that from $1s^2\ 2s^2\ 2p_x^{\ 1}\ 2p_y^{\ 1}$? In hybridizing carbon's orbitals, the valence electrons go from $2s^2\ 2p_x^{\ 1}\ 2p_y^{\ 1}$ to $2s^1\ 2p_x^{\ 1}\ 2p_y^{\ 1}\ 2p_z^{\ 1}$. Then they hybridize to form 4 equal orbitals of $2sp^3\ 2sp^3\ 2sp^3\ 2sp^3$. The number 2 comes from the highest principle quantum number 2, and sp^3 means that one s orbital and three p orbitals were combined and divided into 4 separate equal orbitals with 1 electron each. These four $2sp^3$ orbitals can combine with the $1s^1$ orbitals of each of 4 H atoms to make methane CH_4. We know the beginning of the hybridization process and the ending. The in between is a model that if it did it that way it would turn out the way it should.

Now consider the molecule ethylene $H_2C=CH_2$ also written as C_2H_4. Ethylene is formed by fruit decomposing on a tree. The ethylene causes other fruit in the same or nearby trees to ripen. The hybridization here is a bit different.

$$2s^2\ 2p_x^1\ 2p_y^1 \rightarrow 2s^1\ 2p_x^1\ 2p_y^1\ 2p_z^1 \rightarrow 2sp^2\ 2sp^2\ 2sp^2\ 2p^1.$$

In this case, 3 orbitals are formed from one 2s orbital and two p orbitals. 3 instead of 4 orbitals hybridized. In ethylene, the sp^2 orbitals form covalent bonds between the C atoms and with the 2 H atoms. These are called s (sigma) bonds. The p orbitals overlap above and below the C atoms. These are called p (pi) bonds. It might seem like there are 2 bonds, one above and one below the plane of the carbon atoms. But that is not the case. They are 1 bond because they are formed by 1 pair of orbitals. In an earlier lesson, the point was made that the orbitals do not show the positions of the electrons but rather their energy levels. The diagram below shows where the electrons are most likely to be found. The bond can appear to have 2 parts (above and below the plane of the C atoms). What appears to be 2 parts is really 1 covalent bond.

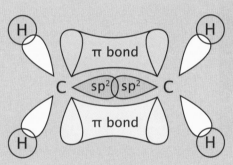

The p orbitals of the C atoms are shared forming 1 p (pi) bond. The bond is above and below the plane of the C atoms. If you think of the C atoms as imbedded in a piece of plywood, the plywood is the plane of the C atoms. The 2 parts of the p bond are above and below the C atoms and the plywood.

Between the C atoms a sp^2 orbital from each C atom is shared to form a s (sigma) covalent bond. The other 2 sp^2 orbitals of each C atom form covalent (s) bonds with the H atoms. Its final structure can be drawn as ...

$$
\begin{array}{c}
H \diagdown \qquad \diagup H \\
\quad C = C \\
H \diagup \qquad \diagdown H
\end{array}
$$

In summary, the π bond is divided into 2 parts above and below the plane of the C atoms and the 3 σ bonds are between the C atoms and between the C atoms and the H atoms.

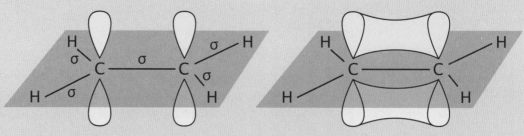

Remember that the orbitals give the energy levels of the electrons rather than their actual positions. The orbital diagrams used to form the bonds with the C atoms are not the actual positions of the electrons but rather where they are more likely to be found. That is why the π bonds can be shown to be partially above the plane of the C atoms and below the plane of the C atoms.

Now consider the molecule acetylene where there are 3 bonds (2π and 1 σ) between the C atoms.

$$H - C \equiv C - H$$

A double bond is stronger than a single bond and a triple bond is stronger than a double bond. When acetylene burns it gets hot enough to melt most metals. When the triple bond is broken, the tremendous energy that holds the C atoms together is released in acetylene torches used in welding.

In this case the hybridization of C is

$$2s^2\ 2p_x^1\ 2p_y^1 \rightarrow 2s^1\ 2p_x^1\ 2p_y^1\ 2p_z^1 \rightarrow sp\ sp\ p\ p.$$

you get 2 sp orbitals and 2 p orbitals. The acetylene molecule looks like ...

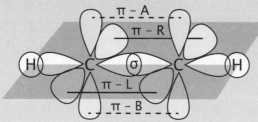

π – R 1/2 of π bond to the right of C atoms
π – A 1/2 of π bond above the C atoms
π – B 1/2 of π bond below the C atoms
π – L 1/2 of π bond to the left of C atoms

The sp orbitals form a σ bond between the C atoms and between the C atoms and the H atoms. The p orbitals form 2 π bonds above and below the plane of the C atoms (1 bond) and to the right and left between the C atoms (the other bond).

What is the point of all this? These detailed properties of C atoms makes possible the many C compounds in what we call organic chemistry and biochemistry. This is what God came up with when He spoke the Universe and life into existence. An amazing thought is that we are told in Scripture that Christ holds all things together for matter to exist. This includes all of these bonds between C atoms.

"And He is before all things, and in Him all things hold together."
Colossians 1:17

It is true that the orbitals are our model and could explain the properties of C atoms. But if you consider the wonderful creativity of God, the real nature of the bonds of C atoms that make up the bulk of living matter may be beyond our capacity to imagine let alone understand. When God finished each day of creation, He said that it was good. What He calls good must be good indeed.

Why do apples ripen on a tree? When the first apples ripen, they produce ethylene.

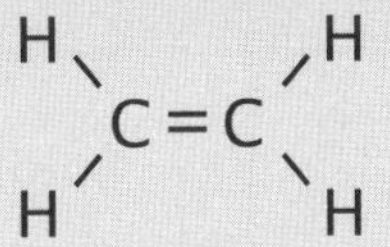

$$\begin{array}{ccc} H & & H \\ \diagdown & & \diagup \\ & C = C & \\ \diagup & & \diagdown \\ H & & H \end{array}$$

The ethylene induces the rest of the apples on the tree and nearby trees to ripen. We like the apples when they are partially ripened (not rotted). But the tree cannot release the seeds from the apples until they complete the ripening process (which we call rotting). The soft squishy rotting apple is producing a lot of ethylene. Have you heard the expression "one rotten apple spoils the whole bunch"? One rotting apple in a barrel of apples produces ethylene that ripens and spoils the rest of the apples.

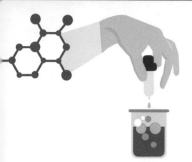

MODELS OF CARBON COMPOUNDS

REQUIRED MATERIALS

- Molecular model kit

PURPOSE

This exercise is to demonstrate the large numbers of different compounds that can be made with carbon because it can form four covalent bonds.

PROCEDURE

1. Make Single-Carbon Molecules:

    A. From the set of molecular models select 1 of the balls that represent carbon atoms.

    B. Using the others for hydrogen atoms and oxygen atoms, make as many different molecules using only 1 carbon atom as you can.

    The number of hydrogen and oxygen atoms that you use are limited by the number of bonds that carbon can make.

2. Now take 2 carbon atoms and see how many molecules you can make.

3. Do it again using 3 carbon atoms.

The carbon atoms each form 4 covalent bonds (you can use double or triple bonds); the oxygen atoms each form 2 covalent bonds and the hydrogen atoms each form 1 covalent bond.

How many molecules did you make using 1 carbon atom, 2 carbon atoms, and 3 carbon atoms?

Describe in your own words the advantage of carbon being able to form 4 covalent bonds. This accounts for the vast number of different molecules that form to make up living cells. This is an excellent example of design.

CHAPTER 25

ORGANIC CHEMISTRY

OBJECTIVES AND VOCABULARY

The Learning Objectives are for the students to gain an understanding (as evidenced by their performance on the quiz) of:

1. Hydrocarbons including alkanes, alkenes, and alkynes

2. The identification of common organic functional groups

3. The isomeric forms of glucose, fructose, and galactose

4. The difference between a mechanistic view and a spiritual view of the nature of organic compounds.

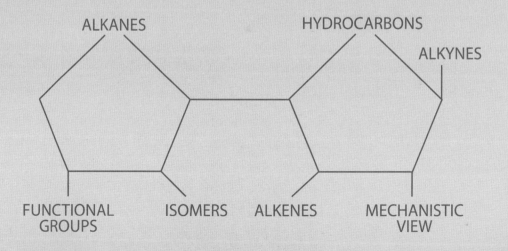

ALKANES   HYDROCARBONS   ALKYNES

FUNCTIONAL GROUPS   ISOMERS   ALKENES   MECHANISTIC VIEW

Termites release about 165 million tons of methane and 55 million tons of CO_2 into the atmosphere every year.

Through hybridization a carbon atom can form 4 covalent bonds. An important exception is carbon monoxide, CO, where the carbon atom forms 2 covalent bonds. CO attaches to hemoglobin molecules in red blood cells, blocking oxygen from being carried to the body's tissues, producing suffocation even though breathing is still possible.

Compounds with just carbon and hydrogen are **hydrocarbons**. Those that have single bonds between the carbon atoms are **alkanes**. The simplest is methane CH_4, also called natural gas, which is used to heat our homes and water and to cook our food. Methane is also a greenhouse gas, meaning that along with other gases it acts like a greenhouse. Ultraviolet light from the Sun heats the surface of the Earth. As the Earth heats up, it gives off infrared radiation in the form of heat. The greenhouse gases trap the infrared radiation so that it cannot escape back out to space. It is called a greenhouse effect, named after a building with transparent walls, such as glass or plastic, used for growing plants in the winter. The plastic or glass traps the infrared radiation heating up the inside of the building. This is the same thing that happens in a car with the windows rolled up in the daylight hours. Termites release about 165 million tons of methane and 55 million tons of CO_2 (another greenhouse gas) into the atmosphere every year. It is popular to think that humans are releasing much more CO_2 into the atmosphere causing global warming. It is true that we are producing some of it but consider the termites and other insects and animals, including cows. Volcanoes also emit a lot of CH_4 and CO_2. It is nice for our egos to think that we are the sole cause of this phenomenon but that is not quite the case. Earth is experiencing a gradual warming, but if considered from a Biblical perspective, the Earth has been warming as a result of the thawing of the ice age that formed after the flood of Noah's day, a relatively short time ago.

Ethane is a 2 carbon alkane ($H_3C–CH_3$). Propane, a chain of 3 carbon atoms with single covalent bonds between them ($CH_3–CH_2–CH_3$). The first carbon atom has a single covalent bond to the middle carbon atom and 3 covalent bonds to H atoms, giving a total of 4 bonds to the carbon atom. The middle carbon atom is bonded to the first and third carbon atoms, giving 2 covalent bonds. It is also bonded to 2 H atoms, giving it a total of 4 bonds. A chain of 4 carbon atoms with single covalent bonds between them is butane that is used in lighters. Each of the carbon atoms is covalently bonded to H atoms, giving each one 4 covalent bonds.

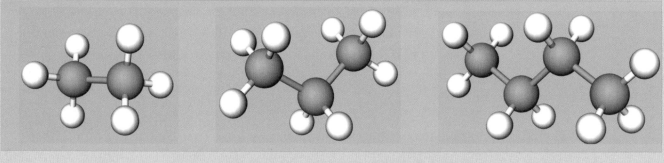

ETHANE   PROPANE   BUTANE

Some of the hydrocarbons have 2 covalent bonds between some of the carbon atoms. These are the **alkenes**. Ethylene ($CH_2 = CH_2$) is an alkene studied in lesson 24. Some hydrocarbons have carbon atoms with 3 covalent bonds between them. These are the **alkynes**. In all of these, H atoms are bonded to carbon atoms to give each one 4 covalent bonds. A common alkyne is acetylene used in acetylene torches for welding.

$$HC \equiv CH$$

Many other organic compounds are named by groups of atoms (called **functional groups**) attached to them. Below are listed some of the major groups of organic compounds with the functional groups for which they are named. A suffix is added to the end of the name of each compound to indicate its functional group.

Group Name	Functional Group	Ending	Example
Alcohol	$- OH$	-ol	Ethanol CH_3CH_2OH
Aldehyde	$\begin{array}{c} H \\ | \\ - C = O \end{array}$	-al	Ethanal $\begin{array}{c} H \\ | \\ CH_3 - C = O \end{array}$
Ketone	$\begin{array}{c} O \\ || \\ - C - C - C - \\ | \\ H \end{array}$	-one	Propanone $\begin{array}{c} O \\ || \\ CH_3 - C - CH_3 \end{array}$
Organic acid	$\begin{array}{c} OH \\ / \\ - C = O \end{array}$	-oic or -ic	Acetic acid $\begin{array}{c} OH \\ / \\ CH_3 - C = O \end{array}$
Ester	$\begin{array}{c} O \\ || \\ - C - O - C - \end{array}$	-ate	Methyl acetate $\begin{array}{c} O \\ || \\ H_3C - C - O - CH_3 \end{array}$
Amine	$- NH_2$	-amine	Ethylamine $CH_3CH_2NH_2$
Ether	$- C - O - C -$	-ether	Dimethyl ether $CH_3 - O - CH_3$

Study the structures of the 6-carbon sugar molecules glucose, fructose, and galactose on the following page.

Which is more like glucose — fructose or galactose? Notice that the galactose is almost identical to glucose except that the 4th carbon down from the top on galactose has its – OH and – H on opposite sides. Enzymes can readily tell the difference between these 2 molecules because the enzymes recognize the shapes of molecules. All 3 of these sugars have the same chemical formula $C_6H_{12}O_6$. They are **isomers** of each other, meaning that they have the same kinds and numbers of atoms but different arrangements. Glucose, also called dextrose, is the product of the breakdown of starches and complex carbohydrates in cells. Glucose is the form of sugar that is metabolized for energy release.

```
        H                         H                         H
        |                         |                         |
        C=O                 H — C — OH                       C=O
        |                         |                         |
  H — C — OH                       C=O                 H — C — OH
        |                         |                         |
 HO — C — H                 HO — C — H                 HO — C — H
        |                         |                         |
  H — C — OH                 H — C — OH                HO — C — H
        |                         |                         |
  H — C — OH                 H — C — OH                 H — C — OH
        |                         |                         |
  H — C — OH                 H — C — OH                 H — C — OH
        |                         |                         |
        H                         H                         H
```

| GLUCOSE | FRUCTOSE | GALACTOSE |

Sucrose, table sugar, is a 12 carbon sugar molecule made up of a glucose molecule combined with a fructose molecule. Fructose is the sugar found in fruit and is the sweetest tasting of the monosaccharides (6 carbon sugar molecules). Galactose is found in brain and nerve tissue. A glucose molecule combined with a galactose molecule form the lactose molecule found in milk.

Early chemists thought that organic compounds could only come from living organisms because that is where they were always found. They mixed chemistry and philosophy and came up with an idea called the vital force. This was the idea that there is a vital (living) force in all living tissue that produced organic compounds. This is not to be confused with the spiritual nature of man taught in the Scriptures.

In 1828, a German scientist Friedrich Wöhler discovered that when the salt ammonium cyanate was heated, urea (found in urine) was produced.

$$NH_4OCN \rightarrow CO(NH_2)_2$$

He produced an organic compound outside of a living organism without a vital force. This led to the **mechanistic view** that many hold to today. This is the idea that everything is physical and that life is just a complex arrangement of molecules and that life can be studied strictly on a

physical level. This denies any spiritual aspect of life. When we realize that the soul that we have from God is not the same thing as the idea of the vital force, we see that Wöhler's work had nothing to do with spiritual reality. We have a physical body that obeys the laws of chemistry but we also have a spiritual nature that can commune with God on a spiritual level. Since Wöhler's time, many other organic compounds have been synthesized and studied outside of living organisms. Many of the synthetics that we have come to rely upon, like polyesters and nylon, are organic compounds. However, it must be strongly noted that life cannot come from non-life. The principle of spontaneous generation, later called "abiogenesis" that tried to explain the rise of life without God has never been shown as anything but an empty hope of evolutionists.

Friedrich Wöhler Monument, Göttingen, Germany

If life consists of only chemical reactions, it would be impossible to explain Jesus' resurrected body. When He died, His cells would have undergone a process called apoptosis where they would be enzymatically digested from the inside. When He met with His disciples after the resurrection …

> As they were talking about these things, Jesus himself stood among them, and said to them, "Peace to you!" But they were startled and frightened and thought they saw a spirit. And he said to them, "Why are you troubled, and why do doubts arise in your hearts?" See my hands and my feet, that it is I myself. Touch me and see. For a spirit does not have flesh and bones as you see that I have." And when he had said this, he showed them his hands and his feet. And while they still disbelieved for joy and were marveling, he said to them, "Have you anything here to eat?" They gave Him a piece of broiled fish, and he took it and ate before them (Luke 24:36–43).

Earlier in Emmaus …

> When He was at table with them, he took the bread and blessed and broke it and gave it to them. And their eyes were opened, and they recognized him (Luke 24:30-31).

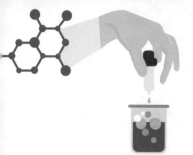

FAT AND WATER SOLUBLE COMPOUNDS

REQUIRED MATERIALS

- Beaker (100 ml), small jar, or glass

- Oil (olive/vegetable)

- Individual types of vitamins, such as A, D, E, K, B, and C (optional)

- Mortar/pestle, blender, or other grinder for solid vitamins

- Other random solutes (see examples in Step 4)

INTRODUCTION

A practical application of the polarities of molecules is their solubilities in water and oils (fats, lipids). Your cells depend upon vitamins, which are molecules that become part of enzymes that catalyze many very important reactions. The enzymes with their accompanying vitamins enable many reactions to occur but are not consumed themselves. Because an enzyme can catalyze many repetitions of a reaction, you do not need them as abundantly as food molecules. Some enzymes are water soluble (polar) and are lost as water leaves the body so they have to be replaced daily. Fat soluble (non-polar) vitamins are stored in fat storage cells and can be retained for long periods of time.

PURPOSE

This exercise demonstrates the concept that "like dissolves like." Polar (water-soluble) molecules dissolve in polar solvents and non-polar (fat-soluble) molecules dissolve in non-polar solvents. This demonstrates the roles of water-soluble vitamins and fat-soluble vitamins.

PROCEDURE

1. Fill a beaker, small jar, or glass about 1/4 full of water.

2. Pour an equal amount of olive oil (or other cooking oil) over the water.

    The oil has a lower density than water so it floats above the water.

3. If you have some vitamins that are not part of mixtures of vitamins (such as vitamin D or vitamin C), place them in the water-oil mixture and notice whether they dissolve in the water or the oil.

    If the vitamin is solid, grind it up into a fine powder. If it is in a capsule, open the capsule and pour out the contents.

Which vitamins dissolved in water? Which dissolved in the oil? Typically, the vitamins A, D, E, and K are fat soluble, and vitamins B and C are water soluble. Does this match your results? In your report, describe your procedure and results.

4. Take several available things and determine if they are water or fat soluble.

    Some examples could be fruit juices, milk, ground-up aspirin, egg whites, blended egg yolk, ketchup, salad dressing, soy sauce, etc. Use fresh containers with oil and water if necessary. Describe what you tested and your results.

CHAPTER 26

BIOCHEMISTRY

OBJECTIVES AND VOCABULARY

The Learning Objectives are for the students to gain an understanding (as evidenced by their performance on the quiz) of:

1. An overview of the role of DNA, RNA in coding for protein construction for controlling life processes

2. The nature of nucleic acids

3. The nature of proteins

4. The nature of carbohydrates and lipids.

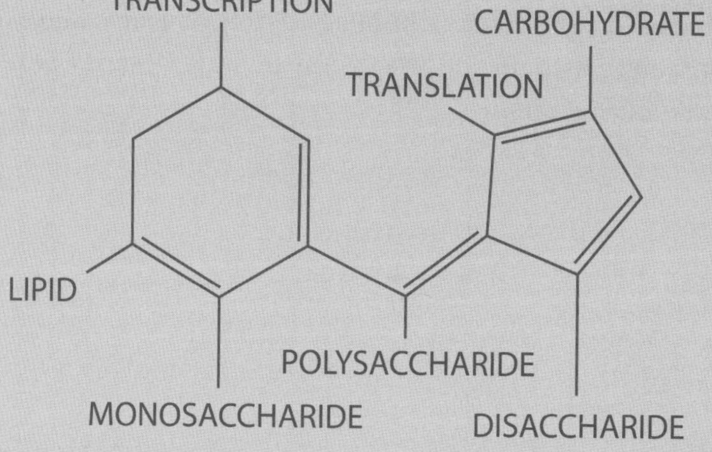

TRANSCRIPTION

CARBOHYDRATE

TRANSLATION

LIPID

POLYSACCHARIDE

MONOSACCHARIDE

DISACCHARIDE

This lesson is not a comprehensive study of the intricacies of biochemistry. This could take a lifetime of study. The chemistry of biological life is characterized by intricate interactions at the molecular level. But life itself is more than molecules. God breathes life into His creatures and removes it. The details of biochemistry cannot be appreciated unless the interactions are taken into account first.

The story begins with DNA (deoxyribonucleic acid). DNA and RNA (ribonucleic acid) are nucleic acids. DNA was first discovered in the nuclei of pus cells, hence the name nucleic acid. They are made up of smaller units called nucleotides. A nucleotide is made up of a 5-carbon carbohydrate (deoxyribose for DNA and ribose for RNA). Attached to this sugar is a phosphoric acid and a nitrogen-containing organic base.

They are linked together into long chains by the phosphoric acid molecules.

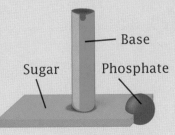

Base

Sugar

Phosphate

Nucleotide

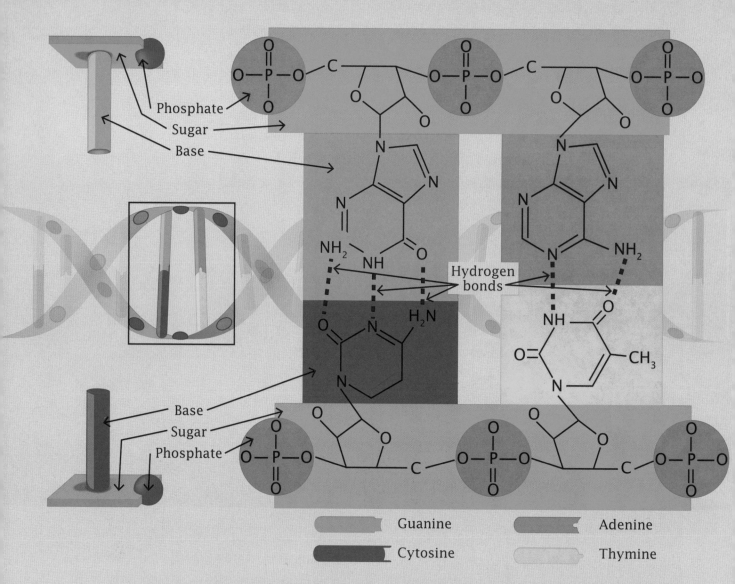

Guanine	Adenine
Cytosine	Thymine

RNA is single stranded. DNA is double stranded with the bases of one strand facing those of the other strand so that the base thymine is opposite the base adenine on the other strand and cytosine is opposite guanine.

The bases of opposite strands face each other so that hydrogen bonds form between the bases. Hydrogen bonds were studied in an earlier lesson dealing with the hydrogen bonds between water molecules. The nitrogen and oxygen atoms of one base form hydrogen bonds with the hydrogen atoms of the opposing base. Guanine and cytosine form 3 hydrogen bonds between each other and thymine and adenine form 2 hydrogen bonds between each other. Like between water molecules, the hydrogen bonds are very weak. But also like water, the hydrogen bonds are in such great quantity to be strong enough to hold the DNA strands together.

In the nucleus of a living cell, a section of the DNA (called a gene) unravels and a copy (a single strand of RNA) of one of the DNA strands is made. This new strand has the sugar ribose instead of deoxyribose in its nucleotides. This process is called **transcription** because a copy is made in the same language (nucleic acid language). To transcribe means to write out a copy.

This shorter strand of RNA is called messenger RNA (m-RNA). It leaves the nucleus of the cell and goes out into the cytoplasm (material of a cell outside of the nucleus), where it is used by enzymes to direct the formation of a protein. Proteins are long chains of amino acids (not mean old acids). The genetic code is such that every sequence of 3 bases in the m-RNA indicates which amino acid is placed next in sequence in a protein. The protein is first made as a long sequence of amino acids. Then other enzymes fold the amino acid chain into a complex 3-dimensional functional protein. The process of constructing a protein from the genetic code in the RNA is called **translation** because it is taking the language of 3 bases of the nucleotides of RNA and translating it into the language of an amino acid sequence.

Deoxyribose
used in DNA backbone

Ribose
used in RNA backbone

Proteins are long chains of amino acids, such as this one.

They are joined together by covalent (called peptide) bonds with the amino- group of one amino acid bonded to the carboxyl (the acid radical -COOH from an organic acid) of the next amino acid.

peptide bond

The amino acid chain is folded into a complex structure under the direction of enzymes (that are also proteins formed by this same process). The completed enzymes have cavities in their structures called active sites into which substrate (molecules that the enzyme is acting upon) molecules fit. This causes the enzyme to change shape, which changes the substrate molecules into product molecules.

> *There are different enzymes for each of the many steps of the many reactions that take place in a cell. They all have to be in place from the start for the cell to function. This is called design.*

There are different enzymes for each of the many steps of the many reactions that take place in a cell. They all have to be in place from the start for the cell to function. This is called design. When the first cells were spoken into being, the enzymes had to all be in place from the very start. The genetic code in the DNA already had to be there with the enzymes ready for transcription and translation to take place. The amino acids had to be available, ready to go. There is not a hint of evolution here.

Simple starch

Cellulose

The word **carbohydrate** means hydrated carbon or carbon and water. The isomers of the 6 carbon sugars glucose, fructose and galactose were studied in the previous lesson. These are called **monosaccharides** because they are the basic units for more complex carbohydrates.

Two monosaccharides combine to form a 12 carbon **disaccharide** molecule such as sucrose (table sugar), which is a glucose combined with a fructose.

Many monosaccharides form complex **polysaccharides,** such as the high energy storage starches. The polysaccharide cellulose cannot be digested except by a few bacterial forms. It forms the fiber part of our diet that cleans out our digestive tract and regulates how fast materials pass through our digestion. They facilitate the removal of toxic and decomposing matter.

Cotton fibers represent the purest natural form of cellulose, containing more than 90 percent of this polysaccharide.

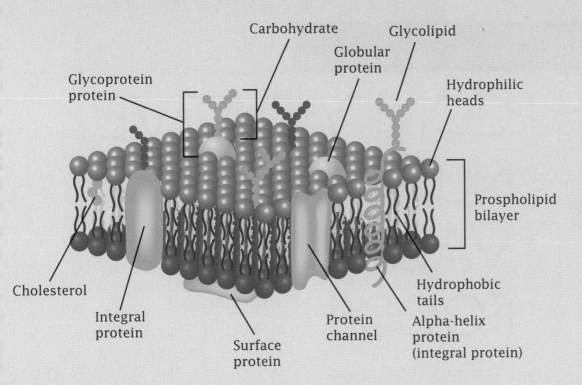

Carbohydrate

Glycolipid

Glycoprotein protein

Globular protein

Hydrophilic heads

Cholesterol

Integral protein

Surface protein

Protein channel

Hydrophobic tails

Alpha-helix protein (integral protein)

Prospholipid bilayer

Lipid (fat) molecules are made up of fatty acids. Fatty acids make up a good part of the plasma membranes surrounding our cells. When we consume more energy in our food than we need, the excess is first stored in the form of starches in our liver and muscle tissue. When our starch reserves are full, the extra then goes into lipids. Lipids are stored in specialized cells called lipid cells. We need a certain amount of lipid but too much can become a health hazard.

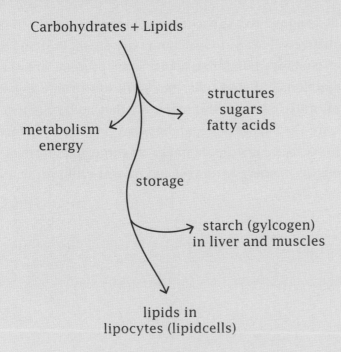

Carbohydrates + Lipids

structures
sugars
fatty acids

metabolism
energy

storage

starch (gylcogen)
in liver and muscles

lipids in
lipocytes (lipidcells)

LABORATORY 26

COMPARATIVE NUTRITIVE VALUES OF FOODS

REQUIRED MATERIALS

• At least 10 different foods with nutrition labels

PURPOSE

This exercise is designed to apply the concepts presented in the lessons to practical nutritive values of foods commonly eaten. Proteins are very large molecules composed of amino acids. Some proteins are enzymes that catalyze reactions in living cells, and some are structural parts of cell membranes, muscle tissue, and materials that hold cells together. Carbohydrates are the sugars and starches that provide and store energy for the living processes. Fats (lipids) make up important parts of cell membranes, long-term energy storage, and insulation to protect internal organs and retain heat.

PROCEDURE

Take at least 10 different foods. Read the nutritive values of them from their labels. Record and compare them to each other. For example, chicken noodle soup has 3 g of protein, 1 g of sugars, 2 g of fat, 890 mg (0.89 g) of Na (sodium), and 50 mg of K (potassium).

Record your data in a chart as shown below (see Teacher Guide):

Food	Protein (g)	Sugars (g)	Fat (g)	Na (g)	K (g)

Which food has the most protein?

Which food has the most sugar?

Which food has the most fat?
The fat content in food usually contains fat soluble vitamins.

Which food has the most Na?
It may be best to avoid it if you are on a low sodium diet.

Which food has the most K?
Some people do not get enough K in their diet. It is needed for nerve impulses, muscle contractions, and cell functions.

CHAPTER 27

RATES OF CHEMICAL REACTIONS

OBJECTIVES AND VOCABULARY

The Learning Objectives are for the students to gain an understanding (as evidenced by their performance on the quiz) of:

1. What is necessary for a chemical reaction to occur

2. The nature of exothermic reactions

3. The nature of endothermic reactions

4. The role of a catalyst in lowering activation energy of a chemical reaction

5. The role of an enzyme as a catalyst in metabolism.

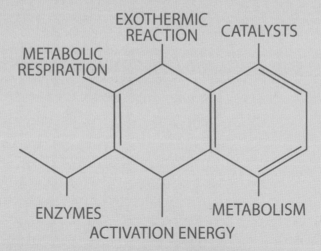

METABOLIC
RESPIRATION

EXOTHERMIC
REACTION

CATALYSTS

ENZYMES

ACTIVATION ENERGY

METABOLISM

I n an earlier lesson it was brought out that a spontaneous chemical reaction was one that would occur with no indication as to how long it would take. This lesson deals with what affects how long it takes for a reaction to occur.

When molecules react with each other, they have to come in contact. To gain an appreciation of what is involved, get with a friend or sibling and 2 lumps of clay. Both of you throw the lumps of clay into the air toward each other, trying to get them to stick together. Do this outside. You will soon discover that this is easier said than done. To make it more realistic, blindfold each of you and try it again. When molecules come together and form new molecules, they are moving and have to collide so that the right part of each molecule hits each other at the right velocity. As you see with the clay, it is a wonder that any chemical reactions ever occur. Molecules are not being directed as you directed the lumps of clay. This is why reactions are more successful when the reactants are more concentrated and the temperature is higher (they move faster). When they are more concentrated, they collide more often because there are more of them. One of the many reasons that there is biological life is that God designed very complex enzymes to capture the reactant molecules and put them together so that they react very quickly. Otherwise, the reactions would be far too slow and biological life would never exist.

Consider this reaction.

$$2 \text{ BrNO (g)} \longleftrightarrow 2 \text{ NO (g)} + \text{Br}_2 \text{ (g)}$$

The Br – N bond must be broken when the BrNO molecules collide with each other and the Br (bromine) atoms must come close enough to each other to form Br_2. This is why spontaneous in chemistry does not mean instantaneous.

When reactants go to form products, they first form a transition state and afterwards form the products. It is similar to when 2 cars crash into each other, they are reactants before they collide. Momentarily both cars are stuck together (this is the temporary transition state). After they bounce apart, they are the products. There is more energy in the transition state than in the individual product molecules separately. That is like saying that there is more energy in the cars smashed together than in each separate crunched car afterward. The following diagram shows the energy relationships of a chemical reaction.

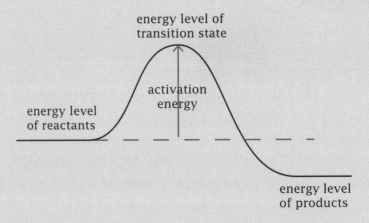

The above diagram represents an **exothermic reaction,** meaning that the products have less energy than the reactants. The product line is lower than the reactant line. Notice that the transition state has more energy than the reactants or the products. Where does the transition state get the extra energy? It came from the velocities of the reactant molecules. When you start a fire in a fireplace or campfire, you have to add energy from another source to get the fire going. This is the **activation energy**. Usually it is from some kind of starter material or kindling. Burning is oxidation, carbon compounds react with oxygen removing electrons from what is burning. Eventually, CO_2 and H_2O are formed. The flames are the energy given off at frequencies we can see by electrons going from higher energy orbitals in the carbon to lower energy orbitals in the oxygen. When

the transition state comes apart, becoming product molecules, the extra energy is given off as heat and possibly light.

A good example of an exothermic reaction is the reaction of HCl (hydrochloric acid) with NaOH (sodium hydroxide). The reaction releases a lot of heat energy. The electrons in the covalent bonds in HCl and NaOH have more energy than the product molecules NaCl (an ionic compound) and H_2O.

$$HCl + NaOH \rightarrow NaCl + H_2O$$

If the products have more energy than the reactants, it is an endothermic reaction. The product line is higher than the reactant line in the following diagram. Heat energy comes from the surroundings.

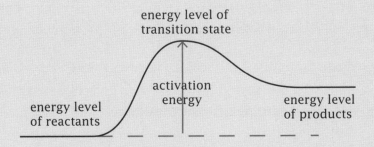

The breakdown of $CaCO_3$ (calcium carbonate) is an example of an endothermic reaction. It occurs more readily as the temperature is raised because the products require more energy than the reactants.

$$CaCO_3 \text{ (s)} \leftrightarrow CaO \text{ (s)} + CO_2 \text{ (g)}$$

You use an endothermic reaction when you use an ice pack that you squeeze to mix the reactants. When the reactants combine, they draw energy from the surroundings and whatever is around the ice pack gets cold.

When the activation energy is high, the likelihood of a reaction occurring diminishes. This is why the rate of a reaction depends upon the temperature. The chemical reactions that occur within our cells have to occur rapidly with very small concentrations at body temperature. The activation energies of the reactions would make them impossible without help. This is accomplished with **catalysts** called **enzymes**. A catalyst is a molecule or atom that is able to substantially lower the activation energy of a reaction. They physically recognize the shape of reactants, attract them with positive and negative charges, and position them so that they come in contact with each other; thus, they can react and form the transition state and then the products.

When the activation energy is high, the likelihood of a reaction occurring diminishes. This is why the rate of a reaction depends upon the temperature.

This speeds up the reactions so that they occur in thousandths of a second rather than hours or days. There are many different enzymes in our cells carrying out many different reactions. They control respiration — use of O_2 and food molecules to release energy — thus keeping our endothermic reactions going and keeping us alive. The breakdown of molecules in our cells to release energy with exothermic reactions is called catabolism. The use of energy to build up other molecules and enable processes to occur, such as nerve impulses and muscle contractions, with endothermic reactions is called anabolism. Together all of the chemical reactions in our bodies are called **metabolism**.

One of the unique properties of a catalyst is that it is not consumed. It helps the reaction but is neither a reactant nor a product. Therefore, you do not have to have them in very large quantities. When it is time for that reaction to stop, another enzyme takes the reacting enzyme apart. The amino acids released from the breakdown of the enzyme are used to make other enzymes.

When food molecules are broken down and energy is released, some of the energy goes into endothermic reactions to build up other molecules, but the majority of the energy is released as heat to maintain body temperature or produce a fever to combat an infection. According to the laws of physics, our bodies are not very efficient because we lose so much energy as heat. But in this case greater efficiency is not good because we would die from low body temperature. God knows what He is doing.

The exothermic reaction for the breakdown of food releasing energy in cells can be summarized as …

$$C_6H_{12}O_6 \text{ (glucose)} + 6\ O_2 \rightarrow 6\ CO_2 + 6\ H_2O + \text{Energy}$$

This is a summary of cellular or **metabolic respiration**. This process has many steps not shown here. Food molecules, other than glucose, are converted into glucose or other molecules that are part of cellular respiration. Lipids, which are also part of our diet, have long chains of carbon atoms. Carbon atoms are removed from a lipid two at a time and enter metabolic pathways releasing energy. If we eat more than we need, God has designed enzymes to connect the excess glucose molecules end to end, making long chains of animal starch called glycogen. Glycogen is stored in liver and muscle tissue. For many of us, our starch storage is full, so further excess is converted into lipid (polite way of saying fat) molecules and deposited in cells in our bodies where we least want it. Perhaps this is part of God's sense of humor to get us to want to eat more wisely.

As a review of several concepts, how many grams of water are produced by the metabolism of 10 grams of glucose ($C_6H_{12}O_6$)?

$$C_6H_{12}O_6 + 6\ O_2 \rightarrow 6\ CO_2 + 6\ H_2O$$

First, we can say that 6 moles of H_2O are produced when 1 mole of $C_6H_{12}O_6$ is metabolized. This is very important for an animal like a desert tortoise whose only source of water is the breakdown of glucose.

The molecular mass of glucose is 6 x C + 12 x H + 6 x O = 6 x 12 + 12 x 1 + 6 x 16 = 72 + 12 + 96 = 180 grams/mole.

$$
\begin{array}{lll}
6\ C & = 6 \times 12\ \text{g/mole} = & 72 \\
12\ H & = 12 \times 1\ \text{g/mole} = & 12 \\
6\ O & = 6 \times 16\ \text{g/mole} = & 96 \\
\hline
C_6H_{12}O_6 & = & 180\ \text{g/mole}
\end{array}
$$

10 g/(180 g/mole) = 0.06 mole of $C_6H_{12}O_6$

$$\frac{10\ \text{grams}}{180\ \dfrac{\text{gram}}{\text{mole}}} = 0.06\ \text{mole}$$

There are 6 moles of H_2O produced from 1 mole of $C_6H_{12}O_6$6.

$$0.06\ \text{mole} \times 6 = 0.36\ \text{mole of}\ H_2O$$

The molecular mass of H_2O = 2 x H + 1 x O = 2 + 16 = 18 g/mole.

$$
\begin{array}{lll}
2\ H & = 2 \times 1\ \text{g/mole} = & 2 \\
1\ O & = 1 \times 16\ \text{g/mole} = & 16 \\
\hline
H_2O & = & 18\ \text{g/mole}
\end{array}
$$

$$(0.36\ \text{mole}) \times 18\ \text{g/mole} = 6.5\ \text{g of}\ H_2O$$

If the tortoise gets enough water from its food, it never has to drink a drop of water (which is often the case). This is God's grace!

Desert tortoise in Snow Canyon State Park, Utah

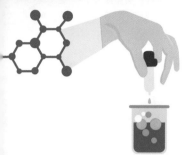

REACTIONS WITH AND WITHOUT CATALYSTS

REQUIRED MATERIALS

- Lugol's solution

- Corn (or other) starch

- Graduated cylinder

- Beaker (250 ml)

- Beaker (100 ml)

- Bunsen/alcohol burner or hot plate

- Stirring rod

- pH paper

- Barium hydroxide

- Vinegar

- Distilled water

- Laboratory scoop/teaspoon measure

- Eyedropper

PURPOSE

This exercise demonstrates the impact of catalysts in lowering the activation energies and reaction times of chemical reactions. Lower reaction times are essential for chemical reactions in living organisms; otherwise, they would be so slow that cells would die.

PROCEDURE

1. Prepare a dilute starch solution.

    A. Place 100 ml of distilled water in a 250 ml beaker and bring it to a boil on a burner or hot plate.

    B. Mix 1 g of soluble starch (corn starch) and several ml of distilled water into a smooth paste in a 100 ml beaker.

    C. Once the water is boiling, carefully remove the beaker from the heat. Mix the paste into the boiling water with a stirring rod, stirring until the starch is dissolved. Your resulting solution may be somewhat cloudy.

    D. Allow the solution to cool to room temperature before using it in experiments.

    A starch solution is not viable for very long: it is recommended to create a fresh batch the day of any experiment requiring it.

2. Take a sample of about 10 ml of the dilute starch solution.

3. Add 1 drop of Lugol's solution (I_2, iodine) to the starch.

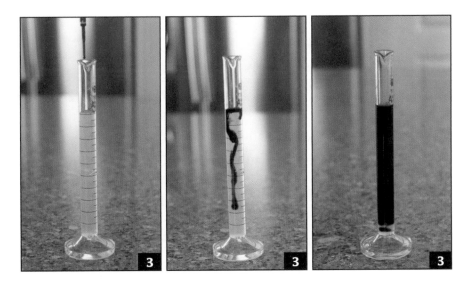

The dark color is an indication of the presence of starch. When the color is darker, there is more starch.

4. Prepare a test sample of saliva mixture.

A. Place a sample of saliva into a teaspoon.

B. Determine the pH of the saliva using the pH paper.

C. Pour the saliva with 10 ml of distilled water into a beaker.

D. Add a couple of drops of starch solution to the saliva in the beaker.

E. Allow the saliva and starch to mix for about 5 minutes.

F. Add a drop of Lugol's solution to the saliva-starch mixture.

   What is the color? What does this tell you about the starch?

G. Allow the saliva and starch to set for another 5 minutes.

   What happens to the color?

5. Prepare a control (unaltered) sample.

A. Take a sample of water about the same volume as the saliva mixture that you used.

B. Determine the pH of the water.

C. Adjust the pH of the water to match that of the saliva sample.

   Prepare some of the Ba (OH)$_2$ solution from lab #22 if you need to raise the pH (make it more basic) or vinegar (acetic acid) if you need to lower the pH (make it more acidic).

D. Add a couple drops of starch to the water, which is at the same pH as the saliva.

E. Allow the water and starch to set for 5 minutes as you did the saliva solution.

F. Add a drop of Lugol's solution to the water-starch mixture.

G. Record its color.

H. Let it set for another 5 minutes and record its color.

What is the difference between the test sample and the control sample? The saliva contains an enzyme called amylase. From your observations, describe the function of amylase. Why is it essential for our lives?

The energy value of food is measured in kJ (kilo Joules). There are 4.18 kJ in a calorie. Starch is a carbohydrate, which yields an average 17 kJ per gram. How many calories is that?

$$17 \text{ kJ} \quad \text{x} \quad \frac{1 \text{ Calorie}}{4.18 \text{ kJ}} \quad = \quad 4 \text{ Calories}$$

How many calories could you get from 25 grams of starch?

$$\frac{4 \text{ Cal}}{g} \quad \text{x} \quad 25 \text{ grams} \quad = \quad 100 \text{ Calories}$$

How many kJ is that?

$$\frac{4.18 \text{ kJ}}{\text{Cal}} \quad \text{x} \quad 100 \text{ Cal} \quad = \quad 418 \text{ kJ}$$

CHAPTER 28

ENVIRONMENTAL CHEMISTRY

OBJECTIVES AND VOCABULARY

The Learning Objectives are for the students to gain an understanding (as evidenced by their performance on the quiz) of:

1. The nature of air pollution

2. The nature of acid rain

3. The nature of catalytic converters

4. The nature of greenhouse gases

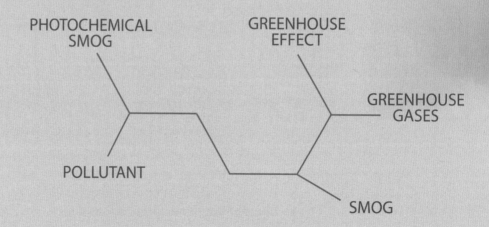

One of the objectives of this study is to be able to read and understand literature that applies concepts of chemistry. This lesson is applied chemistry. Rather than introducing new concepts, this lesson is applying concepts already introduced.

Environmental chemistry is a very complex varied subject. The first topic that most think of is pollution. A **pollutant** is a contaminant or impurity. An example is an oil spill in the ocean. The oil in an automobile is a good thing but not in the ocean and along the seashore. One pollutant that we are all aware of is **smog**. This is a combination of smoke and fog. But the components of smog in different situations can be very harmful. Sometimes we wonder why God created something that is so harmful. The answer lies in how we handle it and how we produce it. One of the blessings of chemistry has been the discovery of ways to cut down on the substances in the atmosphere and water ways that should not be there.

During a thermal inversion that occurs in a valley surrounded by mountains, a layer of cooler air acts like a lid over warmer air. The trapped warmer air contains pollutants that are not able to be

dispersed by wind. This was even noticed in the early days of the Los Angeles basin when smoke from Indian campfires would be trapped by an inversion layer.

Two types of smog form in major cities – gray air and brown air. During cold wet winters, gray industrial smog develops from the burning of coal and other fossil fuels for the production of heat, electricity, and transportation. Dust, smoke, soot, ashes, asbestos, oil, lead, and other heavy metals, along with sulfur oxides, are released into the air, at times in lethal concentrations. In 1952, about 4,000 people in London died from gray air pollution. There are still major problems in China, India, Mexico, and other developing countries. Some of the problems are being helped by technologies to filter the exhaust air before it is released.

In 1952, about 4,000 people in London died from gray air pollution.

In warm climates, a brown haze or **photochemical smog** develops and is trapped by inversion layers. This is an example of brown air pollution. It concentrates in areas with natural basins like Los Angeles and Mexico City. The chief culprit is nitric oxide (NO) that is converted into nitrogen dioxide (NO_2). Sunlight causes NO_2 to react with hydrocarbons (from partially burned fuels), forming photochemical oxidants. The word photochemical means that light is involved in a reaction.

Oxidants in the air include ozone (O_3) and peroxyacyl nitrates called PANS. A small amount of PANS causes stinging of the eyes, lung irritation, crop damage, and corrosion of rubber products. Ozone is invisible and tasteless. You can feel like the air is very clean but have a major smog alert and suffer respiratory damage from ozone in the air. But in our higher atmosphere, ozone is a good thing because it shields us from excessive ultraviolet radiation from the sun. Areas of increased population growth have more unhealthy smog days.

Rain and snow normally have a pH of about 5.7 because of carbonic acid formed by CO_2 combining with water molecules. Combustion and industrial processes can release nitrogen oxides that react with oxygen to form nitric acid (HNO_3) and sulfur oxides that form sulfuric acid (H_2SO_4) producing acid rain, which has been measured to have a pH as low as 2.1. In dry weather, fine particles containing oxides briefly stay airborne but eventually fall to the ground as dry acid deposition that increases the acid levels of bodies of water and the surfaces of buildings, statues, automobiles, etc.

As part of the exhaust system of an automobile, a catalytic converter oxidizes CO and unburned hydrocarbons, converting them to CO_2 and H_2O. As well, they reduce nitrogen oxides to atmospheric N_2. The inside of the catalytic converter has layers of aluminum oxide (Al_2O_3) impregnated with transition metal oxides, as well as platinum, palladium, and rhodium. The metal catalysts absorb O_2 from the exhaust to oxidize CO to CO_2 and oxidize hydrocarbons to CO_2 and H_2O. NO and NO_2 are reduced to N_2. The N_2 is harmless in the atmosphere.

When ultraviolet radiation from the sun passes through the atmosphere, it heats up the ground. The warmer ground emits more infrared radiation (heat) that should pass into space to balance the Earth's temperature. **Greenhouse gases** reflect infrared radiation back to Earth, raising the temperature of the atmosphere. These gases include O_2, CO_2, and NO_2. The energy levels of visible light go from red light with the lowest energy to orange, yellow, green, blue, to violet with the highest energy. Ultraviolet (UV) is invisible and has energy greater (ultra- means greater) than violet. Infrared (IR) is invisible and has less energy (infra- means less) than red light. Most of the UV from the sun is reflected back into space by the ozone in the upper atmosphere. The UV that does reach the surface of the Earth provides heat. Objects whose temperatures

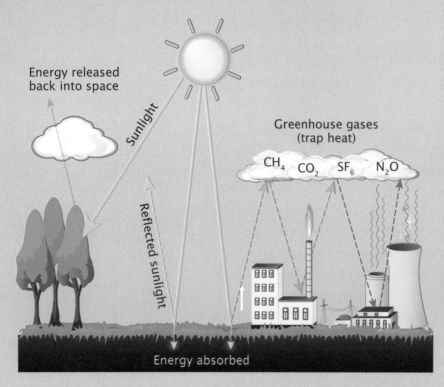

Energy released back into space

Sunlight

Reflected sunlight

Greenhouse gases (trap heat)

CH_4 CO_2 SF_6 N_2O

Energy absorbed

are increased give off IR as heat. Greenhouses that house plants in the winter are made from glass or clear plastic that allows the UV to penetrate but traps the IR and heats up the contents of the greenhouse. This is the **greenhouse effect**. Probably the planet with the greatest greenhouse effect is Venus. Its atmospheric pressure is about 100 times greater than ours and is about 90 percent CO_2. This is very different than Earth. Some claim that if we keep emitting CO_2, we will become like Venus. But we could not come close to that.

There is some data supporting global warming in some areas of Earth. It has not occurred to many that this warming could be from quite a different source than humans producing greenhouse gases. It is very likely that there was a major ice age after the flood of Noah. Since then we know that large masses of glaciers have melted back. If they had not, about half of the North American continent would be uninhabitable. That would be a greater problem. What we see today could just be part of that phenomenon rather than a result of human caused pollution. Nevertheless, it is to our benefit to reduce harmful pollution. Another explanation that some scientists have proposed is that the global warming is part of a normal cycle that the Earth goes through and that we just have not been able to measure it long enough to see the whole picture. There are many natural sources of CO_2 and nitrates in the atmosphere that we do not cause or control — such as volcanoes. Remember too that CO_2 is necessary in the atmosphere for the photosynthesis of plants. The other greenhouse gas is also very necessary — oxygen. It is interesting that O_2 is far more abundant in the atmosphere than CO_2. In fact CO_2 is a rare gas in our atmosphere. So, if we were going to make a major reduction of greenhouse gases, we would have to reduce the O_2. Ouch!

What we see today could just be part of that phenomenon rather than a result of human caused pollution. Nevertheless, it is to our benefit to reduce harmful pollution.

Alternative energy sources are being developed that do not require the burning of fossil fuels. A major difficulty is that the research that is required beforehand is very expensive. The expenses are great before any substantial financial benefit is realized. Some alternative fuels being used today are methane (CH_4, natural gas) and ethanol.

This last lesson is an example of applied science. Pure science is the discovery of principles and data that can be applied to solve problems and improve our overall life style. The procedures developed in the early days of research into the immune system are paying off today in the development of many diverse vaccines. God is gracious in providing these tools of grace as we live in a fallen world.

In studying chemistry at this time, you may be wondering what you will do with this information and insights. That is a question that only the Lord can answer. Whether you go on studying chemistry in college or not, this course will give you insights in handling many everyday situations as you go through life. As well, as you later raise a family, you may be home schooling your children through high school.

Remember as well that this study is our trying to, by God's grace, understand the world around us, keeping in mind that God has given us His wisdom through His Word. We can build upon our understandings from the Scriptures, and if we pay attention to Him, it can help keep us from many errors from our own understandings. Science is good but subservient to God and His Word.

The planet that demonstrates the greatest greenhouse effect is Venus.

LABORATORY 28

SOIL TESTING

REQUIRED MATERIALS

- Soil testing kit

- 4 soil samples

INTRODUCTION

This exercise is a practical application of many of the principles that you have studied throughout this course. Soil pH is important in successfully growing plants. It seems that some people are very successful at gardening and agriculture and others just give up after repeated failures. A lot of it has to do with the pH of the soil. Some test their soil and work with it to get the pH where it should be and others are fortunate in that what they want to grow matches the native soil conditions.

Remember that pH of 7 is neutral (neither acidic nor basic), pH below 7 is acidic, and pH above 7 is basic. Plants whose veins in the leaves are parallel (corn, lilies) and evergreens do better in acidic soil. Dark green leaf plants whose veins branch out usually do better in basic soil.

PURPOSE

A standard soil testing kit tests soil for pH, nitrogen (NO_3^-, nitrate), K (potassium), and P (phosphorus). Plants acquire C (carbon) and O (oxygen) from CO_2 in the atmosphere. These are necessary to form carbohydrates from photosynthesis. In order to produce proteins, phospholipids, nucleic acids, and other necessary compounds, they require N and P from the soil. K is necessary for many cell processes. This lab is a practical exercise in using a soil testing kit and applying the results.

PROCEDURE

Follow the instructions in the kit testing for pH, N, P, and K. Choose at least 4 different soil samples. Choose them from diverse sources, such as exposed ground, soil under the lawn, fertilized soil, etc. Prepare a chart for your report similar to the following (see Teacher Guide):

Soil Sample Location	pH	N	P	K

Describe what is growing or what you would like to grow in each location. Soil kits usually list which plants grow best under varying conditions. What should you do to condition the soils for the best growing conditions?

Location # 1

What do you want to grow in this soil?

Do you need to condition the soil?

How would you condition it?

Location # 2

What do you want to grow in this soil?

Do you need to condition the soil?

How would you condition it?

Location # 3

What do you want to grow in this soil?

Do you need to condition the soil?

How would you condition it?

Location # 4

What do you want to grow in this soil?

Do you need to condition the soil?

How would you condition it?

God created plants with their unique traits and needs just like He made us. An important part of chemistry is understanding the ways to use the resources that our loving Father has given to us.

Acidic soil

Basic soil

CONCLUSION

After this study, it should be apparent that the creation is one of order rather than chaos. Chaos means random processes with no purpose or guiding principles. In physical chemistry, order is defined as predictability. If you have perfume molecules in a bottle, you know where they are — in the bottle. If you take the cap off the bottle, they diffuse out and make the whole area smell like perfume. The perfume molecules

were ordered in the bottle because you had a better idea as to where to locate them. Once they escaped the bottle, their disorder (your ability to find them) increased greatly. Some may even escape out into earth's atmosphere and even out to space beyond earth.

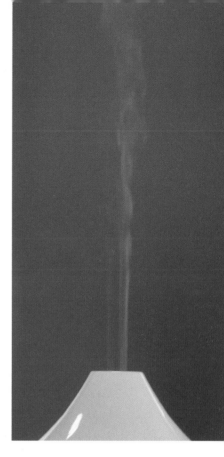

Some have claimed that diffusion, which is the most pervasive physical phenomenon in the universe, demonstrates lack of order and purpose in the universe. It is true that we cannot predict where to find an individual perfume molecule, but we can predict where to find them in bulk. In 1846, Thomas Graham demonstrated that at constant temperature the effusion (gas escaping through a small hole) rate of gases was proportionate to the square root of their molecular mass. In other words, smaller molecules spread out in less time than larger molecules. This means that the rate at which gas molecules spread out is a mathematical relationship. This became known as Graham's Law. It takes longer for heavier molecules to spread out. This means that diffusion is also subject to the constraints of other physical processes. Our inability to predict the motions and locations of individual perfume molecules is due more to our lack of understanding rather than a lack of purpose or direction.

Some say that God is only involved in the big issues of life, but He is even involved in the movement of perfume molecules.

What do we learn about God and ourselves from chemistry? We learn that God's ways are always the same; He is sovereign; He controls the finest details of the universe; He demonstrates His grace (even in a sin-cursed universe the physical processes operate consistently); and He made provision for us in a world where we can function and live. Remember that there is a big difference between the physical realities of chemistry and how people choose to interpret them. He is our Creator God and Savior.

> *Thus says the LORD, who gives the sun for light by day and the fixed order of the moon and the stars for light by night, who stirs up the sea so that its waves roar — the LORD of hosts is his name: "If this fixed order departs from before me, declares the LORD, then shall the offspring of Israel cease from being a nation before me forever" (Jeremiah 31:35–36).*

> *He is the image of the invisible God, the firstborn of all creation. For by him all things were created, in heaven and on earth, visible and invisible, whether thrones or dominions or rulers or authorities — all things were created through him and for him. And he is before all things, and in him all things hold together (Colossians 1:15–17).*

LABORATORY PROCEDURES

INTRODUCTION

All of the labs are written assuming an understanding of basic laboratory procedures and practices. This may not be the case for all students, though, so this section contains an overview of practical laboratory methods.

WEIGHING

Weighing chemicals in a lab is a simple procedure as long as a few basic rules are followed:

1. Calibrate your scale before use (if you have a calibration weight).

2. Always tare your weigh boat before weighing your chemicals. Taring is the process of placing your weigh boat on the scale and pressing the tare (or zero) button on your scale so that the mass of the weigh boat is not included with the chemicals you are measuring. If your scale does not have a tare function, it is recommended that you record the mass of the weigh boat after the scale has started up. Turning on the scale with the weigh boat on it can throw off your calibration. The process is as follows:

   A. Turn on scale (wait for it to show zero)

   B. Place weigh paper / weighing boat (container) on the scale

   C. Tare the scale, or record the mass of the container

   D. Add the chemical you are measuring until you reach the mass desired (total mass including the recorded mass of the container if you did not tare).

3. Use weigh paper or weighing boats to keep chemicals contained.

    A. Weigh paper is generally easier to use for small quantities of dry chemicals and may be creased at 1/3 of its width to help guide dry chemicals into narrower mouthed containers (small beakers / test tubes).

    B. Weigh boats are stiffer and hold a larger amount more easily, but dry chemicals can stick to them, making precise measurement more difficult. If measuring dry chemicals for mixing into a solvent, you may transfer a small amount of the solvent into the boat, using a pipette and pipette the solution back into the sample of solvent.

USING A MULTIMETER

Multimeters vary: the simplest are capable of measuring AC and DC voltage (in volts, designated by a *V* symbol), current (in amps, designated by an *A*), and resistance (in ohms, designated by Ω). There are different scales for each quantity measured, and a good general practice when measuring an unknown quantity is to start at the highest end of the scale for the quantity being measured and lower the setting incrementally until getting a reading. The actual process of taking a reading is straightforward:

1. Connect the probes to the appropriate ports on the multimeter.

2. Set your dial to the quantity that you're measuring on the smallest number that is larger than your expected input value (i.e., to 20V if you're testing a 9V battery). When testing an unknown value, start at the high end and work your way down through the range while testing.

3. Touch the tips of the probes to the contacts for the device you're measuring.

PIPETTING

Pipetting is the technique used to transfer liquid reagents between holding containers. There are several tips for improving your technique and accuracy when measuring volumes being transferred.

1. **Fit your pipette tip carefully when using a 2-piece pipette (pipette tip and pipette pump).**
   A poor or forced fit can damage the pipette seal and affect your results.

2. **Pre-wet and flush your pipette tip before any transfers.**
Aspirate (fill the pipette) with your solution and empty it into a suitable waste container several times to wet the sides and flush out any residual substances.

3. **Aspirate at 90 degrees, and dispense at 45 degrees.**
Filling with your pipette perpendicular to the solution ensures that it fills correctly and that you can get an accurate reading off the side. Dispensing at 45 degrees ensures that the liquid fully dispenses without leaving residue in the tip.

4. **Aspirate with the tip 2–3 mm below the surface.**
This ensures that no air is being aspirated into the pipette and that you release a small amount back into the solution to get the desired volume.

5. **Release the plunger slowly.**
While filling the pipette, it is important to release the plunger slowly to avoid aspirating air bubbles.

6. **Dispense against the side of the receptacle.**
This goes hand-in-hand with dispensing at 45 degrees: dispense against the side of the receptacle or into the liquid already in the receptacle. Never shoot or squirt the liquid into the air: this can affect your measurement results.

7. **Touch off at the very end of your receptacle.**
This means to touch the droplet at the tip of your pipette to the side of the beaker or the surface of your solution, allowing surface tension to "pull" the last droplet off of the tip.

8. **Don't hold the pipette when not actively transferring solution.**
Holding a pipette not in use allows for thermal transfer, which can affect liquid density and air pressure in the pipette, changing your measurements.

BURETTE PREPARATION / TITRATION

A burette is used to precisely measure the volume of a liquid added: the scale starts at the top with 0, and the volume of solution delivered is calculated by subtracting the initial reading from the final reading.

1. Rinse the burette with distilled water

   A. Fill the burette with distilled water.

   B. Drain a few ml through the stopcock (valve or bead) to flush the tip.

   C. Invert the burette over a sink to drain the remainder of the water (opening the stopcock if necessary to help drain quicker).

2. If there are drops clinging to the inside of the burette:

   A. Add a small amount of detergent and clean the inside of the burette with a long brush.

   B. Rinse and flush three times with distilled water.

      i. The first two rinses may be emptied out the top.

      ii. The final rinse should be drained completely through the stopcock to flush out the tip.

3. Prepare your solution in the burette:

   A. Fill the burette with 5–10 ml of your solution.

   B. Tilt and roll the burette to coat the inner surface with your solution, and drain the solution through the stopcock. Discard this waste.

   C. Fill the burette *above* the zero line and drain the excess into a waste beaker until the bottom of the meniscus rests on the zero line. This will flush any remaining air bubbles out of the stopcock and delivery tip.

      It is not necessary to have the initial reading exactly on zero, but it makes the calculation of the difference in readings much easier.

4. Titrate your burette solution into the sample flask by controlling the stopcock (or bead) with one hand and gently swirling your flask with the other to mix the reagents.

   Be very careful not to overshoot by adding liquid too quickly: allow the mixture time to stabilize between additions from the burette as you approach equilibrium.

There are many varieties of heat sources that may be used in the lab, but ideally you will have access to either a Bunsen burner or gas stove. Otherwise, an alcohol burner may be used.

> **WARNING: Anytime you are working around open flame there is a risk of clothing or other objects catching fire. Take precautions to stay safe, such as wearing clothing that fits snugly, not wearing long jewelry, and always wearing safety goggles.**

1. Bunsen Burner:

    A. Connect your burner to the gas supply.

    B. Close the needle valve (the valve on the bottom of the burner).

    C. Turn on the gas supply.

    D. Open the needle valve ½ turn.

    E. Light the burner by taking your ignited match or lighter and running it up the side of the barrel until the gas ignites. If using a spark lighter, place the cup ½ inch above the top of the burner and spark several times until the flame ignites.

    This prevents a tall flame from burning your hand as it might if you held the lit match over the top of the burner)

    F. Adjust the flame until you have a clear blue flame surrounding an inner blue cone.

    There should be no yellow edges to a properly adjusted flame.

2. Alcohol Burner:

    A. Using the purest alcohol you have access to (91 percent isopropyl rubbing alcohol is sufficient but does not burn as cleanly as a Bunsen burner on gas), fill the alcohol burner's reservoir.

    B. Light the wick / burner following the directions for your burner.

    C. If your wick is adjustable, adjust until you have a minimum of orange or yellow flame.

3. When using your burner to heat samples, remember that the hottest portion of the flame is the tip of the inner blue cone.

APPENDIX 2

Can a person be a Christian and a chemist? The answer is yes! The following scientists contributed to the area of chemistry and honored God with their lives. The first list is a chronological history of those scientists who have passed away while the second list shows the names of contemporary scientists of the faith who have contributed to the field of chemistry.

CHRONOLOGICAL HISTORY OF CHRISTIAN CHEMISTS

Credit for the discovery of arsenic as an element is generally given to **Albertus Magnus** (1193–1280), a medieval Germanic theologian, scholar, and scientist. Despite being a deadly poison in large doses, arsenic is an essential trace element (i.e., at very low concentrations) for many animals and perhaps for humans.[1]

Albertus Magnus
(1193–1280)

Robert Boyle (1627–1691) one of the founders of the Royal Society of London, is generally credited with being the father of modern chemistry, as distinct from the alchemy of the Middle Ages. His name is associated with the basic law he discovered relating gas pressures to temperature and volume, the fundamental principle of gas dynamics. His contributions in both physics and chemistry are very great in number, and he was considered to be probably the greatest physical scientist of his generation.

Yet he was also a humble, witnessing Christian and a diligent student of the Bible. He was profoundly interested in missions and devoted much of his own money to Bible translation work and the propagation of the gospel. He was strong in apologetics, founding via his will the "Boyle lectures" for proving the Christian religion.[2]

Robert Boyle
(1627–1691)

1 Duncan, Richard D. *Elements of Faith: Faith Facts and Learning Lessons from the Periodic Table*. Green Forest, AR: Master Books, 2008.
2 Morris, Henry M. *Men of Science, Men of God: Great Scientists Who Believed the Bible*. Green Forest, AR: Master Books, 1982.

Richard Kirwan (1733–1812) was an Irish chemist and mineralogist, president of the Royal Irish Academy for 23 years, and author of the first systematic treatise on mineralogy, also making many contributions to chemistry. He also advocated Flood geology and vigorously opposed the increasingly influential uniformitarian theories of James Hutton, the predecessor of Sir Charles Lyell.[3]

Richard Kirwan
(1733–1812)

Reverend William Gregor (1761–1817) was the pastor in the county of Cornwall, England. While he was a beloved and respected minister to his parishioners, Gregor was a man of many interests, including painting, etching, and music. But his favorite hobby was mineralogy. He had no formal science education but eventually earned quite a reputation in the field. His most lasting discovery was a mineral that he found near his church, a gray-black magnetic sand that looked a lot like gunpowder. After much study, Gregor concluded that this mineral was probably a mixture of iron oxide and the oxide of a new unknown metal.

Later, a German chemist, Martin Heinrich Klaproth, verified that Gregor's mineral did in fact contain a new metal. Klaproth was a respected scientist, but he insisted on giving credit for the element's discovery to the amateur Gregor. However, Klaproth claimed the privilege of naming the new element for his own. He called it "titanium," after the Titans, the ancient giants of Greek mythology.[4]

Reverend William Gregor
(1761–1817)

John Dalton (1766–1844) was born in a Quaker family and was a practicing Quaker all his life, during a time when Quakers were all known as orthodox and pious, Bible-believing Christians. Throughout his life, he was known as a godly man, of very simple tastes and lifestyle.

In science, he is best recognized today as the father of modern atomic theory, which revolutionized the study of chemistry. His first love, however, was meteorology, and he formulated the well-known gas law of partial pressures. He was also the first to recognize and describe the phenomenon of color-blindness, a condition also known ever since as Daltonism. Dalton was one of the founders of the British Association for Advancement of Science, in 1831. One year later he was awarded a doctorate by Oxford University.[5]

John Dalton
(1766–1844)

3   Ibid.
4   Duncan, *Elements of Faith*8.
5   Morris, *Men of Science*.

John Kidd, MD (1775–1851) was professor of chemistry at Oxford during most of his career and made many significant contributions in this field. He pioneered the use of coal as a source of chemicals, his work eventually providing the foundation for the development of synthetics. As a well-respected Christian, he was chosen to present one of the Bridgewater Treatises, entitled *The Adaptation of Nature to the Physical Condition of Man.*[6]

John Kidd, MD
(1775–1851)

Sir Humphry Davy (1778-1829) was a truly amazing man. He was very famous in his time. His brilliant lectures, complete with flashy chemistry demonstrations, were attended by throngs of admirers, kings and queens, and common people from all over Europe. Some of his best friends were the poet Samuel Taylor Coleridge and Dr. Peter Roget, who created the thesaurus. Davy wrote a classic manual on angling that fishermen still refer to. Davy was so highly respected on both sides of the English Channel that Napoleon himself awarded him a prestigious science medal. (Even though England and France were at war at the time!)

Davy's discovery of so many elements and his invention of the mine safety lamp were just a few of Davy's scientific accomplishments. He discovered the anesthetic properties of nitrous oxide (N_2O or laughing gas) and suggested its use in surgery. (Of course, it is still used in dentistry to this day!) ... Humphry Davy remained a devout Christian who always glorified God in his work.[7]

Sir Humphry Davy
(1778-1829

William Prout (1785–1850) authored one of the Bridgewater treatises: *Chemistry, Meteorology, and the Function of Digestion, Considered with Reference to Natural Theology.* As a chemist and physiologist, he was an early leader in the sciences of nutrition and digestion and was the first to identify basic foodstuffs as fats, proteins, and carbohydrates. He is best known, however, for recognizing that the atomic weights of elements could be identified as a series of relative whole numbers.[8]

6 Ibid.
7 Duncan, *Elements of Faith.*
8 Morris, *Men of Science.*

Sir Joseph Henry Gilbert
(1817–1901)

Sir Joseph Henry Gilbert (1817–1901) was one of the prominent fellows of the Royal Society who signed the Scientists' Declaration, affirming his faith in the Bible as the Word of God and opposing Darwinist materialism. As an agricultural chemist, he developed nitrogen and super-phosphate fertilizers for use with crops and helped develop (as first co-director) the world's first agricultural experimental station, located in Hertfordshire in 1843. He also served as professor of rural economy at Oxford University.[9]

Thomas Anderson (1819–1874) was a Fellow of the Royal Society and a prominent Scotch chemist, discoverer of pyridine and other organic bases. As Regius Professor of Chemistry at Glasgow, he also edited the *Edinburgh New Philosophical Journal*. He was one of the signatories of the Scientists' Declaration of 1864, affirming his faith in the scientific accuracy of the Bible and the validity of the Christian faith.[10]

Thomas Anderson
(1819–1874)

Dr. Joseph Lister (1827–1912) is considered the father of modern surgery. He was responsible for many medical advances, including drainage tubes for wounds and "catgut" sutures, which dissolve in the human body. But his most important innovation was the use of "antiseptic techniques." Before the 1870s, surgeons did not wear gloves or gowns. Often they made no attempt to even clean up between surgeries. Many considered it a status symbol to be covered with blood from previous operations. As a result, about 50 percent of patients who had major surgery died from infections, often after otherwise successful operations. In the days before the discovery of germs, doctors believed that infections arose spontaneously and that there was no way to stop them.

By 1860 Lister was already a successful surgeon when he began a lifelong friendship with the French chemist (and fellow Christian) Louis Pasteur. Pasteur's work had shown that the spoiling of milk and wine resulted from germs in the air. Lister concluded that such germs might also cause human infection. If so, there might be ways to prevent germs from reaching his patients. Lister began looking for ways to do just that.

Lister experimented with several chemical solutions he called "antiseptics" in his operating rooms to clean wounds and soak bandages.

Dr. Joseph Lister
(1827–1912)

9 Ibid.
10 Ibid.

He even sprayed them in the air to kill bacteria before they could reach the patient. The results were remarkable. His patients had practically no infections. However, after repeated exposures, these antiseptics were harmful to Lister and the surgeons who worked with him, bleaching and numbing the skin and causing vision and breathing problems. Lister eventually found a chemical that was very effective yet safe for medical personnel; it became a standard antiseptic in operating rooms for decades. This chemical was boric acid (HBO_2).

Like so many of the giants in the history of science, Dr. Lister was a faithful Christian. A contemporary of Lister wrote that he was a humble servant of God who "always asked for His guidance in moments of difficulty. ... (He) believed himself to be directly inspired by God. ... To Lister, the operating theatre was a temple."[11]

In 1894 **William Ramsay** (1852-1916) and his partner, **Lord Rayleigh** (1842-1919), startled the scientific world when they announced his discovery of a new gaseous element, which they called argon. This non-reactive, inert gas had gone unnoticed throughout the whole history of chemistry. All the great minds of science had gone about their discovering and inventing, totally oblivious to this inert gas inhabiting the very air that they breathed. A hidden gas!

William Ramsay
(1852-1916)

Of course Ramsey knew of Mendeleev's periodic table, now 25 years old. He knew that all of the other elements existed in columns of similar elements with different molecular weights. So he became convinced that there were other hidden gasses, similar to argon, waiting to be found.

Just a year later, Ramsey detected a gas that was given off by a radioactive, uranium ore. He collected it and determined that it was yet another hidden gas, lighter than argon but just as inert: helium.

In 1898 Ramsey tried out a new method to find additional inert gasses. This time he had another partner, Morris Travers. They used super-cooled liquid air to separate out various components of the atmosphere based on their boiling points. In the same year, Ramsey and Travers discovered three more inert gases. In order, they were krypton (the "hidden" gas), neon (the "new" gas), and xenon (the "strange" gas). Sir William Ramsey holds a unique place in the history of science. No other person has discovered an entire unknown column of the periodic table. His discoveries, with his co-workers, led to many technical advances and a fuller understanding of the nature of atoms and

Lord Rayleigh
(1842-1919)

---

11 Duncan, *Elements of Faith.*

elements. As Rayleigh, the co-discoverer of argon, put it, they "sought out ... the [hidden] works of the Lord" (referring to Ps. 111:2).

In 1904 William Ramsey and Lord Rayleigh each received Nobel Prizes for their joint discovery of the noble gas, argon. They were life-long friends and both were well known as devout Christians. They believed that in their study of science they were merely uncovering the hidden things of God.[12]

George Washington Carver
(1864–1943)

George Washington Carver (1864–1943) was a great scientist who was considered the world's top authority on peanuts and sweet potatoes and their products.

Born a slave, he worked his way through college in the North and then returned to the South, desiring to devote his life to improving the quality of southern farmlands and the economic prosperity of his people. As a faculty member at the Tuskegee Institute in Alabama, he turned down a number of much more lucrative offers, as the fame of his genius as an agricultural chemist spread. He developed over 300 products from the peanut and over 118 from the sweet potato.

Carver was also a sincere and humble Christian, never hesitating to confess his faith in the God of the Bible and attributing all his success and ability to God. In 1939 he was awarded the Roosevelt medal, with the following citation: "To a scientist humbly seeking the guidance of God and a liberator to men of the white race as well as the black."[13]

Charles Stine
(1882–1954)

Charles Stine (1882–1954) was for many years the director of research for the E.I. duPont company. As an organic chemist with many degrees and honors, he developed many new products and patents for his company. He was a man of top eminence in his field but also a simple believing Christian. He frequently spoke to scientific and university audiences concerning his faith and also wrote a small book entitled A Chemist and His Bible. After a stirring exposition of the gospel and an appeal to accept Christ, Dr. Stine gave this testimony of the Creator: "The world about us, far more intricate than any watch, filled with checks and balances of a hundred varieties, marvelous beyond even the imagination of the most skilled scientific investigator, this beautiful and intricate creation, bears the signature of its Creator, graven in its works."[14]

---

12 Ibid.
13 Morris, *Men of Science.*
14 Ibid.

MODERN CHRISTIAN CHEMISTS

Professor Edward A. Boudreaux is professor emeritus of chemistry at the University of New Orleans, Louisiana. He holds a B.S. in chemistry from Loyola University, an M.S. in chemistry, and a Ph.D. in chemistry from Tulane University. Professor Boudreaux has spent 29 years in graduate education and research in the area of theoretical and inorganic chemistry and chemical physics, and is the author or co-author of four technical books in the area of inorganic chemistry, as well as numerous scientific papers in peer-reviewed journals and textbooks.[15]

Professor D.B. Gower is emeritus professor of steroid biochemistry at the University of London, United Kingdom. He holds a B.S. in chemistry from the University of London, and a Ph.D. in biochemistry from the University of London and was awarded a D.Sc. from the University of London for his research into the biochemical mechanisms for the control of steroid hormone formation. Professor Gower is a fellow of the Royal Society of Chemistry, a fellow of the Institute of Biology, and a chartered chemist.[16]

Dr. Bob Hosken is senior lecturer in food technology at the University of Newcastle, Australia. He holds a B.S. in biochemistry from the University of Western Australia, an M.S. in biochemistry from Monash University, a Ph.D. in biochemistry from the University of Newcastle, and an M.B.A. from the University of Newcastle. Dr. Hosken has published more than 50 research papers in the areas of protein structure and function, food technology, and food product development.[17]

Dr. John K. G. Kramer is a research scientist with Agriculture and Agri-Food Canada. He holds a B.S. (hons) from the University of Manitoba, an M.S. in biochemistry from the University of Manitoba, and a Ph.D. in biochemistry from the University of Minnesota and completed three years of postdoctoral studies as a Hormel fellow at the Hormel Institute and as an NRC fellow at the University of Ottawa. Dr. Kramer has identified, characterized, and synthesized the structure of numerous food, bacterial,

15 Ashton, John F. *In Six Days: Why Fifty Scientists Choose to Believe in Creation.* Green Forest, AR: Master Books, 2007.
16 Ibid.
17 Ibid.

and biological components and has published 128 refereed papers and numerous abstracts and book chapters. He was one of the core scientists who evaluated the toxicological, nutritional, and biochemical properties of canola oil and demonstrated its safety. He presently serves as associate editor of the scientific journal *LIPIDS*.[18]

Dr. John P. Marcus is research officer at the Cooperative Research Centre for Tropical Plant Pathology, University of Queensland, Australia. He holds a B.A. in chemistry from Dordt College, an M.S. in biological chemistry, and a Ph.D. in biological chemistry from the University of Michigan. Dr. Marcus's current research deals with novel antifungal proteins, their corresponding genes, and their application in genetic engineering of crop plants for disease resistance.[19]

Dr. Linda S. Schwab is professor of chemistry at Wells College, USA. She holds a B.A. in chemistry from Wells College and an M.S. and a Ph.D., both in organic chemistry, from the University of Rochester. Dr. Schwab served for four years as associate in the Center for Brain Research, School of Medicine and Dentistry, University of Rochester. She specializes in undergraduate chemistry education and natural product research and has published 18 papers and conference abstracts.[20]

Dr. A.J. White is student advisor, dean of students office, at the University of Cardiff, in the United Kingdom. He holds a B.S. with honors, a Ph.D. in the field of gas kinetics from the University College of Wales, Aberystwyth, and has completed a two-year, post-doctoral fellowship at the same university. Dr. White subsequently served in a number of university administrative posts. Over the years he has written several books and numerous articles relating to creation-evolution and science and the Bible, as well as making several appearances on British television and radio programs dealing with these issues.[21]

18 Ibid.
19 Ibid.
20 Ashton, John F. *On the Seventh Day: 40 Scientists and Academics Explain Why They Believe in God.* Green Forest, AR: Master Books, 2002.
21 Ashton, *In Six Days.*

acid — Something that releases H⁺ ions into water. ... 27, 65, 68, 73, 81–82, 87, 95, 103, 146, 148, 150, 153, 158, 167, 173, 177, 181, 183, 187, 189–192, 197–206, 208–219, 221–227, 229, 244–245, 247–249, 255–256, 260, 262, 265, 281

activation energy — The energy it takes to form an intermediate when reactants become products. It is the energy supplied by kindling wood when you are trying to start a camp fire. A catalyst lowers the activation energy, making the reaction go faster. ... 252, 254–255

alkali metals — Metals in Group 1 of the periodic table. They have 1 electron in their highest energy s orbital. They include Na (sodium) and K (potassium). ... 145, 153, 174, 177–178

alkaline earth metals — Metals in Group 2 of the periodic table. They have 2 electrons in their highest energy s orbital. They include Ca (calcium) and Mg (magnesium). ... 147, 153, 174, 178

alkanes — Hydrocarbons with single covalent bonds between the carbon atoms. ... 236–237

alkenes — Hydrocarbons with at least one double covalent bond between 2 carbon atoms. ... 236, 238

Magnesium, an alkaline earth metal

alkynes — Hydrocarbons with at least one triple covalent bond between 2 carbon atoms. ... 236, 238

alternating current (AC) — Electrons move back and forth through a metal. ... 176

amine — Organic compounds with ⁻NH₂ attached to a carbon atom. ... 239

anion — A negatively charged ion such as Cl⁻. ... 148, 157, 179, 185

atom — The smallest unit of matter that can be identified as an element (such as hydrogen, calcium, and iron). ... 9–12, 14, 16–17, 20, 38, 40, 45–48, 52, 65–70, 77–79, 81, 83, 85–89, 92–93, 95, 99–102, 104–105, 109, 115–121, 123, 128–132, 139, 145–146, 150–151, 156–158, 164–168, 174–178, 180, 185–186, 228–233, 235, 237–239, 246, 254–256

atomic mass — The mass of a mole of atoms of a given element (such as the mass of a mole of hydrogen atoms is 1 gram/mole). ... 36, 38, 44, 47–48, 105, 135

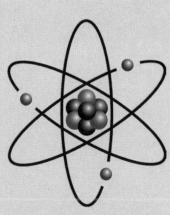

Stylised atom with blue dots as electrons, red dots as protons, and black dots as neutrons.

The study of the chemistry of living cells and organisms is biochemistry.

atomic mass number — Total number of protons and neutrons in an atom that establishes the total mass of an atom. ... 47

atomic mass unit (amu) — The mass of a proton or a neutron. ... 47–48

atomic number — Number of protons in an atom that establishes the identity of an element. ... 46, 48, 119–121, 127, 129, 132, 156–157

Avogadro's number — 6.022×10^{23} ... 36–38

base — Something that releases OH^- into water. ... 64, 68, 82, 167, 177, 190–192, 199, 200–203, 206–212, 214, 216–218, 221, 225–227, 245–247

biochemistry — The study of the chemistry of living cells and organisms. ... 233, 244–245, 283

buffer (pH buffer) — Molecules that absorb added H^+ ions and OH^- ions, keeping the pH constant. ... 220–227

buffering capacity — The amount of molecules available that can absorb added H^+ ions and OH^- ions. ... 220, 222–225

carbohydrates — Compounds of carbon, hydrogen, and oxygen. We normally think of them as sugars and starches. ... 65, 229, 239, 244, 248, 250, 268, 279

Glucose, a type of carbohydrate

catalyst — A substance that lowers the activation energy of a chemical reaction, greatly increasing its reaction rate. ... 66, 252, 255–256

cation — A positively charged ion such as Na^+. ... 148, 157, 179, 185

centi — The prefix in the metric system indicating 1/100 of a unit (a centigram is 1/100 of a gram). ... 18, 21, 24

chemical equation — An equation that describes reactants becoming products in a chemical reaction. A balanced chemical equation shows how many reactant molecules become a definite number of product molecules. ... 69, 77, 85, 92, 115

chemical reaction — The atoms in molecules are rearranged in different combinations. ... 10, 12, 64–69, 72–75, 80, 82, 96, 105, 115–118, 135–136, 138, 209, 215, 221, 241, 252–256, 258

chemistry — The study of matter. ... 8–11, 14–15, 18, 20–22, 26–27, 29, 32, 37–38, 40, 52, 55, 57, 60, 66, 94, 115, 118, 130, 143, 148, 157, 167, 179, 193, 201, 221, 223, 228, 233, 236, 240–241, 245, 254, 262–263, 267, 270–271, 277–281, 283–284

chromatography — Lab techniques set up for the separation of mixtures. ... 52, 55, 110, 113

concentration — The amount of solute divided by the amount of solvent. ... 28, 30, 32–35, 37, 60, 63, 194, 197–198, 202–206, 208, 211–215, 223–224, 227, 255, 264

covalent bonds — Shared electron pairs between non-metal atoms. These are true physical bonds. ... 81, 132, 158, 164, 166–168, 175, 229–231, 234–235, 237–238, 255

deci — The prefix in the metric system indicating 1/10 of a unit (a decigram is 1/10 of a gram). ... 18, 21, 24

density — The mass of an object divided by its volume (such as grams per milliliter). ... 10, 12, 18, 22–23, 26, 169, 243, 274

deuterium — A hydrogen atom with 1 proton and 1 neutron. ... 47

diprotic acid — An acid molecule that releases 2 H$^+$ ions. $H_2SO_4 \rightarrow 2H^+ + SO_4^{--}$. ... 213

diprotic base — A base molecule that gives up 2 OH$^-$ ions. $Ca(OH)_2 \rightarrow Ca^{++} + 2\ OH^-$. ... 213

direct current (DC) — Electrons move in one direction through a metal. ... 176

disaccharides — These are combinations of 2 monosaccharides. A common example is sucrose, which we put on food. ... 248

dissociation constant — The concentrations of the products of a chemical reaction times each other where the concentration of each is raised to the power of the coefficient of each from a balanced equation divided by the reactant concentrations times each other. For example, in the equation $H_2SO_4 \leftrightarrow 2H^+ + SO_4^{--}$, the dissociation constant is $[H^+]^2[SO_4^{--}] / [H_2SO_4]$. ... 194, 212–213

ductile — A property of metals whereby they can be pulled out into thin wires. ... 176

electron configuration — The electron orbitals and numbers of electrons in each orbital of an atom (such as $1s^2\ 2s^2\ 2p_x^2\ 2p_y^1\ 2p_z^1$ for an oxygen atom with 8 electrons). For $2p_x^2$ the $2p_x$ orbital has 2 electrons. ... 114, 119–121, 126–129, 132, 134–136, 138, 141–143, 146–147, 155–158, 165–167, 229–230

electron orbitals — Designations of energy levels of electrons within atoms. ... 114, 123, 127, 130

electronegativity — The tendency to attract electrons. ... 136

A graduated cylinder filled with various liquids illustrates density

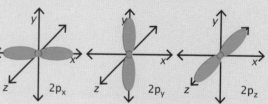

The $2p_x$, $2p_y$, and $2p_z$ orbitals within electrons

electrons — Particles of an atom that are outside of the nucleus of an atom and have a charge of minus one and negligible mass. ... 9, 44–47, 66, 68, 114–121, 123–125, 127–139, 141–143, 145–150, 155–158, 162–163, 165–168, 174–181, 183, 185–187, 189, 199, 215, 229–232, 254–255

electrophoresis — A technique of separating proteins in an electric field. ... 113

element — The identity of an atom such as hydrogen, carbon, or iron. ... 9, 14, 20, 46, 48, 81–82, 86, 92, 95, 101, 105, 107, 116, 119, 120–121, 123, 126, 129, 131, 133–139, 141–147, 149–153, 157, 175, 177–179, 229–230, 277–278, 281

E = mc² — Says that atoms can be converted into pure energy (energy equals the mass of an object times the speed of light squared). ... 90–91

endergonic — Chemical reactions that absorb energy so that the products have more energy than the reactants. ... 64, 67

enzymes — Proteins that catalyze chemical reactions. ... 64, 66–67, 83, 193, 221, 239, 242, 247–248, 250, 252, 254–256, 261

equivalence point — This is reached when all of the reactants in the standard solution have reacted with the reactants in the added solution. ... 205

ester — Organic compounds with – C – O – C – with another oxygen atom double bonded to one of these C atoms. They have odors similar to fruits and perfumes. ... 239

ether — Organic compounds with – C – O – C – in the molecule. These can be toxic and in some cases explosive. ... 239

exergonic — Chemical reactions that give off energy so that the products have less energy than the reactants. ... 64, 67

exothermic reaction — The products have less energy than the reactants. ... 254–256

ferromagnetic — A property of atoms where they behave as tiny magnets like iron atoms. ... 150

First Law of Thermodynamics — Says that energy cannot be created or destroyed. ... 77

functional groups — Groups of atoms attached to carbon atoms in organic molecules. ... 236, 239

galvanic cell — Also called a voltaic cell in which metal atoms of one element force electrons onto metal atoms of another element. ... 185, 188

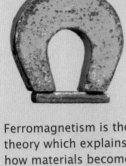

Ferromagnetism is the theory which explains how materials become magnets.

Smog from factories contributes to the greenhouse effect.

gram — The unit of mass in the metric system. ... 21, 27, 30, 33, 38, 47, 61, 105, 197–198, 261

greenhouse effect — Ultra-violet light rays penetrate a planet's atmosphere and warm the surface. The warm surface gives off infra-red radiation, which is trapped by the atmosphere warming the planet. ... 237–238, 266–267

A halogen lightbulb.

greenhouse gases — Reflect infrared radiation back to earth, raising the temperature of the atmosphere. ... 237, 262, 265–266

groups — (see periodic table groups)

halogens — Elements in Group 17 of the periodic table. They tend to gain 1 electron such as $Cl + e^- \rightarrow Cl^-$. ... 82, 148, 153

Henderson-Hasselbalch equation — An equation whereby the concentrations of buffering compounds can be determined as needed to maintain the pH of a solution. ... 224, 227

hybridizing orbitals — When electron orbitals are combined and divided equally as in the case of carbon. This accounts for the s and p orbitals coming out at the same energy level. ... 230

hydrocarbons — Compounds of carbon and hydrogen. ... 236–238, 264–265

hydrogen bonding — Attractions between opposite charges of polar molecules (such as water molecules).

Hydrogen bonds in water

hydronium ion — Acid molecules where $H^+ + H_2O \rightarrow H_3O^+$. ... 191

hydroxyl ion — Base molecules OH^-. ... 168–169

hypothesis — A proposed explanation for observed phenomena. ... 13-15, 23, 128

ion — An atom or molecule that has a + or – charge. ... 47, 68, 103, 120–121, 124–125, 148, 151, 155–159, 162, 168, 177, 179–180, 183,189, 191–194, 198–199, 202–204, 211, 213–214,217–219, 222–223, 226–227

ionic bonds — Attractions between positive metal ions and negative charged non-metal ions. ... 121, 154–155, 157–158, 165–166, 168

isomers — Molecules with the same elements and numbers of atoms of each element but in different arrangements. ... 82–83, 239, 248

isotopes — Atoms that have the same number of protons (same element) but different numbers of neutrons (different masses). ... 44, 47–48

One kilo is 1,000 grams.

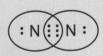

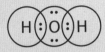

Lewis structures for N_2 and H_2O

ketone — Organic compounds with an oxygen atom double bonded to a carbon atom within a carbon chain. ... 239

kilo — A 1,000 of something in the metric system (a kilogram is 1,000 grams). ... 21, 24, 261

kinetics — Study of the rate of chemical reactions. ... 64, 66, 117

Law — (see scientific law)

Law of Definite Proportions — States that the elements in a compound will always be in the same proportions no matter where they came from. ... 101

Lewis structures — Diagrams of valence electrons using dots around the symbol for the element. They show which valence electrons are paired and which are unpaired. ... 164, 167–168

lipid — These are non-polar fat molecules usually derived from cholesterol. ... 65, 229, 242, 244, 249–250 256

liter — The unit of volume in the metric system. ... 21, 25

malleable — A property of metals whereby they can be pounded out into flat sheets. ... 176

mass — The measure of the amount of matter in an object. ... 10, 12, 15, 21–22, 25–26, 30, 33–34, 36, 38, 41, 44, 47–49, 59, 61–62, 90, 101–102, 104–105, 135, 217, 257, 266, 271–272

mathematical modeling — Describing a system in nature based on mathematical concepts. ... 84, 91

mechanistic view — The view that life is composed of molecules and chemical reactions and nothing more. ... 236, 240

metabolic respiration — Reactions and processes in molecules in our cells to release energy. ... 256

metabolism — All of the chemical reactions that occur within a living cell. ... 47, 64, 66–67, 82–83, 252, 256–257

metal — Elements on the left of the jagged line of the periodic table that tend to give up electrons to non-metal atoms. Metals are good conductors of heat and electric currents. ... 9–10 66, 94–95, 114, 116–117, 120–121, 127–129, 136–137, 145–150, 153, 157–158, 164–165, 174–183, 186, 189, 264–265, 278

metalloids — Elements along the jagged edge of the periodic table with properties between those of metals and non-metals. They are also called semiconductors and used in solid state digital electronics. ... 147, 174, 179

Tin (Sn), a soft and malleable metal, was pressed into ornate ceiling tiles in the early 1900s.

meter — The unit of length in the metric system. ... 21, 24, 162, 202–203, 205

metric system — A system of measurements based upon units of 10. ... 18, 20–22, 24, 94

micro — The prefix in the metric system indicating 1/1,000,000 of a unit (a microgram is 1/1,000,000 of a gram). ... 21

milli — The prefix in the metric system indicating 1/1000 of a unit (a milligram is 1/1000 of a gram). ... 21, 24

Water molecules (H_2O) cover 71 percent of the Earth's surface.

model — A description of the properties and behavior of something that cannot be seen (such as atoms and molecules). ... 10–11, 16–17, 80–82, 92–93, 95, 101, 117, 128, 131–132, 151, 231, 233–235

molarity — The number of moles of a solvent in a liter of solution. ... 36–39, 56, 192, 197, 219

molar concentration — See molarity.

molar solution — Contains 1 mole of solute in 1 liter of the solution. ... 57

mole — 6.022×10^{23} of something (such as carbon atoms) in the same way that a dozen of something is 12. ... 36, 38, 40–41, 43, 48–49, 58–59, 61–62, 97–100, 102, 104, 107–109, 204–205, 217, 219, 257

molecular mass — The mass of a mole of particular molecules (such as the mass of a mole of water molecules is 18 grams/mole). ... 36, 38, 41, 44, 48–49, 59, 61–62, 102, 104, 217, 257, 271

molecule — Two or more atoms bonded together (such as a water molecule H_2O). ... 11–13, 15–17, 20, 37–38, 40, 42, 44, 47–48, 52, 54–55, 65–71, 77–79, 81–83, 85–87, 89–90, 93, 95, 97, 100–101, 104–105, 109, 111, 116, 120, 132, 149, 150, 155, 162, 165–171, 173, 176–177, 191–192, 194, 203–204, 206, 208–209, 213, 217–219, 221, 226, 228–232, 235, 237, 239–240, 242, 245–246, 248–250, 253–256, 265, 270–271

monoprotic acid — An acid molecule that releases 1 H^+ ion. $HCl \rightarrow H^+ + Cl^-$. ... 213

monoprotic base — A base molecule that gives up 1 OH^- ion. $NaOH \rightarrow Na^+ + OH^-$. ... 213

Helium, a noble gas, is used to fill balloons.

monosaccharide — 6 carbon carbohydrate molecules. Common examples are glucose, fructose, and galactose.

neutrons — Particles of an atom that are within the nucleus of an atom and have a zero charge and mass of 1 atomic mass unit. ... 44–47, 199

A pH meter is an easy way to test the pH of liquids.

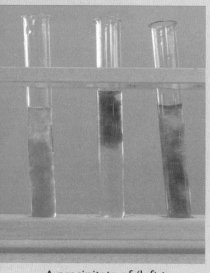

A precipitate of (left to right) copper hydroxide, a precipitate of iron (III) hydroxide, and the third a precipitate of iron (II) hydroxide.

noble gases — Elements with their p orbitals filled so they do not gain nor lose electrons. They are very non-reactive. ... 128–130, 138, 149

non-metal — Elements on the right of the jagged line of the periodic table that tend to take electrons from metal atoms. Non-metals are not good conductors of heat and electric currents. ... 120–121, 129, 145–147, 157–158, 164–165, 175

non-polar molecules — Molecules uniform in charge with no + and – ends. Examples are oils and fats. This is why oil and water do not mix. ... 111, 170–171

nucleic acids — Molecules of DNA and RNA. ... 65, 223, 229, 244–245, 268

octet rule — Atoms are more stable in chemical bonds when they have 8 valence electrons (with the highest principle quantum number). ... 128–129, 137, 147–149, 157–158, 166

organic acid — Organic compounds with the ^-COOH group. ... 247

organic chemistry — The study of carbon compounds. ... 233, 236, 284

oxidation (oxidize) — The removal of electrons from atoms. ... 64, 66, 68–69, 93, 116, 129, 147–149, 174, 177–183, 185–187, 215, 254, 265,

oxidizer — Something that removes electrons from other atoms. ... 147–149, 179

percent concentration — The mass of solute divided by the mass of solvent times 100. ... 28, 30, 32–35, 37

Periodic Table of the Elements — A visual representation of the chemical elements, ordered by atomic number, electron configuration, and chemical properties. ... 48, 115, 118–119, 121, 127, 129, 134–136, 140–145, 151–152, 157, 159, 177–178, 230

periodic table groups — Vertical columns of the periodic table. ... 136, 142, 144–145, 149, 152–153, 159, 178

periodic table periods — Horizontal rows of the periodic table. ... 136

pH — The negative logarithm of the H^+ concentration. If $[H^+] = 10^{-4}$ M, the pH is 4. ... 190, 192–198, 200–203, 205–208, 216–217, 219–227, 258, 260, 265, 268–269, 283–284

photochemical smog — Sunlight causes NO_2 (nitrite) and hydrocarbons in the atmosphere to become harmful compounds. ... 264

photons — Units of light identified by their energy. ... 116–117

polar molecules — Molecules with opposite charges on their ends, such as H_2O molecules. ... 54, 111–112, 155, 168, 170–172, 242, 288, 291

pollutant — Contaminants or impurities. ... 263

polysaccharides — These are combinations of many monosaccharides. Common examples are starch and cellulose. ... 248

precipitation — A process where the product of a chemical reaction comes out of solution and settles to the bottom of the container. ... 64, 67, 69, 214–215

Tiny crystals of salt precipitated from Dead Sea water, magnified 25x

principle quantum number — In the expression $3s^2$, 3 is the principle quantum number indicating the energy level of the orbital. The number 2 is the number of electrons in the 3s orbital. ... 128–129, 137, 157, 166–167, 230

protons — Particles of an atom that are within the nucleus of an atom and have a charge of plus one and mass of 1 atomic mass unit. ... 9, 16, 44–48, 68, 116, 119–121, 129, 146–147, 151, 156, 162, 168, 179, 185, 199, 208–209, 229

quanta (quantum) — Designates a certain amount of energy (such as in $3s^2$, the number 3, principle quantum number, indicates the energy level of the 3s orbital). ... 116–118, 128–129, 133, 137, 139, 157, 166–167, 230

reducer — Something that gives electrons to other atoms. ... 179

reduction (reduced) — Giving electrons to atoms. ... 64, 68–69, 120, 129, 174, 177–183, 185–187, 189, 214–215, 265–266

salt — The combination of a metal ion and a non-metal ion usually formed when an acid reacts with a base (such as NaCl formed when HCl reacts with NaOH). ... 11, 13, 29–30, 32–34, 40–41, 60, 68, 121–122, 125, 148, 152–153, 155, 157, 159–160, 170, 182, 196–197, 203, 214, 224–225, 227, 240

scientific law — Consistent observations, such as the law of gravity. These are observations, not explanations (which would be hypotheses or theories). ... 128

semiconductors — Elements along the jagged edge of the periodic table with properties between those of metals and non-metals. They are also called metalloids and used in solid state digital electronics. ... 147, 179

View of smog south of Los Angeles City Hall, September 2011.

Acid and base titration is a quantitive analysis of the concentration of an unknown acid or base solution.

smog — The combination of smoke and fog. ... 263–264

solubility — How much solute dissolves into a solvent (such as how much salt will dissolve in a liter of water). ... 30, 34–35, 210, 214

solubility product — This is like the dissociation constant, except that you do not divide by the concentrations of the reactants because the reactants are not in solution. For the reaction $Cr(OH)_3 \longleftrightarrow Cr^{+++} + 3\ OH^-$, the Ksp is $[Cr^{+++}][OH^-]^3$. ... 214

solute — Something dissolved in something else, such as a salt (solute) dissolved in water. ... 28–30, 33–34, 36–38, 56–58, 215, 242

solution — A solute dissolved into a solvent, such as salt water. ... 28–39, 48, 56–62, 67, 160–163, 178–179, 182, 191–193, 196–197, 199, 201–208, 210, 212, 214–219, 222, 224–225, 227, 258–260, 273–275

solvent — Something that something else is dissolved into, such as salt dissolved into water (solvent). ... 28-29, 34, 38, 112, 210, 212, 242, 273

spin — A property of electrons in which electrons of opposite spin combine in a single orbital. ... 150

spontaneous — In chemistry, this means that a chemical reaction can occur with no indication as to when it will occur. ... 64, 66, 241, 253–254

strong acid — Acid molecules that release all of their H^+ ion (such as hydrochloric acid HCl). ... 192, 202, 212

strong base — Base molecules that release all of their OH^- ions (such as NaOH). ... 202–203, 217–218

theory — A hypothesis that has been thoroughly tested and supported by many experts. ... 13, 44–45, 117, 128, 133, 139, 278

thermodynamics — The study of the transfer of heat between objects. ... 77

titration (titrate) — A method of measuring the concentration of an acid solution by gradually adding a base until all of the acid is neutralized. ... 200, 202–205, 216, 218–219, 275

transcription — The process whereby mRNA (messenger RNA) is made as a copy of part of a DNA strand. ... 247–248

transition metals — These are metals in Groups 3–12 of the periodic table. They represent the filling of the d orbitals and do not follow the Octet Rule. ... 128, 149–150, 174, 178, 265

translation — The process whereby the code in the mRNA is translated into the amino acid sequence in synthesizing protein molecules. ... 247–248, 277

triprotic acid — An acid molecule that releases 3 H$^+$ ions. $H_3PO_4 \leftrightarrow 3H^+ + PO_4^{-3}$. ... 213

tritium — Hydrogen atom with 1 proton and 2 neutrons with an atomic mass of 3 amu. ... 47

valence electrons — Electrons in an atom with the highest principle quantum number. ... 137, 145, 147, 157, 165, 180, 230

voltage — The force exerted when metal atoms force electrons onto other atoms. This is also called the electromotive force (EMF). ... 179, 182, 186–187, 189, 273

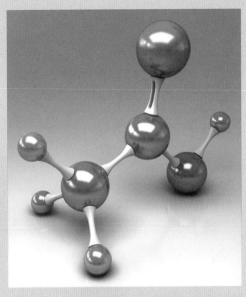

Acetic acid is an example of a weak acid.

voltaic cell — Also called a galvanic cell where metal atoms of one element force electrons onto metal atoms of another element. ... 185, 187

weak acid — Acid molecules that only release some of their H$^+$ ions (such as acetic acid CH_3COOH). ... 192, 203, 212, 217–218, 223–224, 226–227

weak base — Base molecules that do not release all of their OH$^-$ ions (such as NH_3OH). ... 203, 212

weight — The pull of gravity upon the mass of an object. ... 10, 12, 22, 38, 47, 58, 104–105, 107-109, 143, 272

JACOBS' GEOMETRY

A Respected Standard for Teaching Geometry

TEACHER GUIDE
High School | Geometry

SOLUTIONS MANUAL
High School

GEOMETRY
Seeing, Doing, Understanding
Third Edition

DVD INSTRUCTION
GEOMETRY
Seeing, Doing, Understanding
Third Edition

GEOMETRY
Seeing, Doing, Understanding
Third Edition

Harold R. Jacobs

Harold Jacobs' *Geometry* has been an authoritative standard for years, with nearly one million students having learned geometry principles through the text. Now revised with a daily schedule, the text is adaptable for either classroom or homeschool use. With the use of innovative discussions, cartoons, anecdotes, and vivid exercises, students will not only learn but will also find their interest growing with each lesson. The full-color student book focuses on guided discovery to help students develop geometric awareness. Geometry is all around us. Prepare to understand its dynamic influence so much better!

Jacobs' Geometry	978-1-68344-020-8
Solutions Manual	978-1-68344-021-5
Teacher Guide	978-1-68344-022-2
3-BOOK SET	**978-1-68344-036-9**
Geometry DVD	713438-10236-8
3-BOOK / 1-DVD SET	**978-1-68344-037-6**

MASTERBOOKS
CURRICULUM

AVAILABLE AT
MASTERBOOKS.COM   800.999.3777
& OTHER PLACES WHERE FINE BOOKS ARE SOLD.

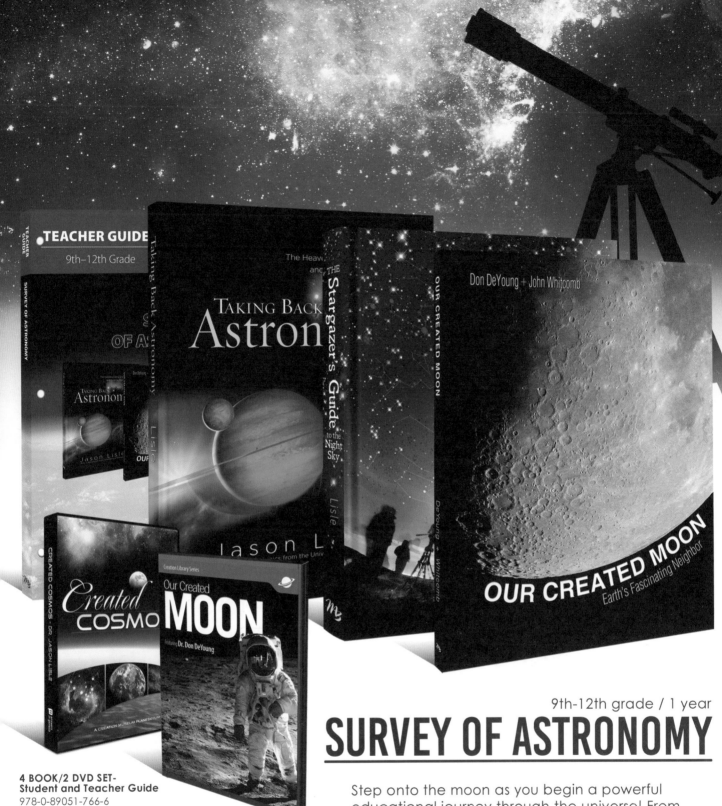

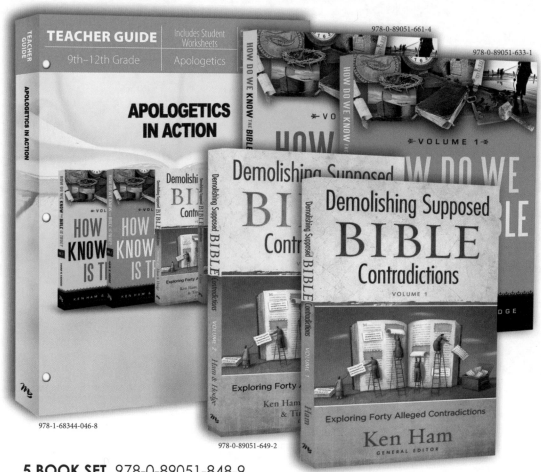

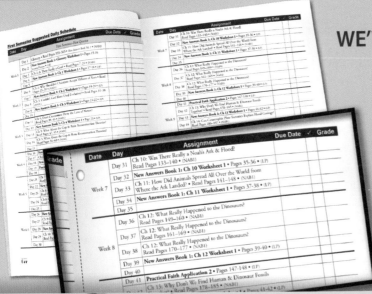